Series on Analysis, Applications and Computation – Vol. 6

An Introduction to Pseudo-Differential Operators

3rd Edition

Series on Analysis, Applications and Computation

Series on Analysis, Applications and Computation – Vol. 6

An Introduction to Pseudo-Differential Operators

3rd Edition

∘ M W Wong

York University, Canada

NEW JERSEY • LONDON • SINGAPORE • BEIJING • SHANGHAI • HONG KONG • TAIPEI • CHENNAI

Published by

World Scientific Publishing Co. Pte. Ltd.

5 Toh Tuck Link, Singapore 596224

USA office: 27 Warren Street, Suite 401-402, Hackensack, NJ 07601

UK office: 57 Shelton Street, Covent Garden, London WC2H 9HE

Library of Congress Cataloging-in-Publication Data
Wong, M. W. (Man Wah), 1951–
    An introduction to pseudo-differential operators / by M.W. Wong (York University,
Canada). -- 3rd edition.
        pages cm. -- (Series on analysis, applications and computation ; vol. 6)
    Includes bibliographical references and index.
    ISBN 978-9814583084 (hardcover : alk. paper)
    1. Pseudodifferential operators. I. Title.
    QA329.7.W658 2014
    515'.7242--dc23

2014003783

British Library Cataloguing-in-Publication Data
A catalogue record for this book is available from the British Library.

Printed in Singapore

Preface

There have been a lot of developments in pseudo-differential operators since the first edition was published in 1991. The second edition, published in 1999, has served well as an introduction to pseudo-differential operators in the capacity of a textbook. The prerequisites for the first two editions are minimal and can be seen from the prefaces for the first and second editions, which follow this one.

The third edition is intended to contain not only improvements of some of the contents and additional exercises to some of the existing chapters in the second edition, but also new chapters to make the book more useful without losing the original intent of keeping the book as elementary as possible. The new chapters notwithstanding, the whole book remains to be a textbook primarily for beginning graduate students in mathematics. It is also useful to mathematicians aspiring to do research in pseudo-differential operators and related topics.

The new chapters are the last seven chapters, Chapters 17–23, of the book. The focus of Chapters 17–21 is on the class of pseudo-differential operators studied in the first two editions. The theme underlying Chapters 17–19 is Gårding's inequality, which is used to prove the existence and uniqueness of solutions of pseudo-differential equations. In particular, the Hille–Yosida–Phillips theorem on one-parameter semigroups is used to prove the existence and uniqueness of solutions of initial value problems for heat equations governed by pseudo-differential operators. After a chapter, Chapter 20, on the general theory of Fredholm operators that we need for this book, the ellipticity and Fredholmness of pseudo-differential operators are developed in Chapter 21. The ellipticity, Fredholmness, an index formula and the spectral invariance for another class of pseudo-differential operators, dubbed symmetrically global pseudo-differential operators in this book, are studied in Chapters 22 and 23. The emphasis of the book, as in the first two editions, is on the global theory of elliptic pseudo-differential operators on $L^p(\mathbb{R}^n)$, $1 < p < \infty$.

As we are now well into the new millennium and moving forward with increasing acceleration, many advanced topics in any area of science and engineering in 1991 are now being taught in basic courses to students. The prerequisites for a complete understanding of the book can be succinctly described as a first course in functional analysis including the Riesz theory of compact operators. The book contains ample material to be studied leisurely and carefully for a two-semester course. One-semester courses can be designed by omitting certain topics in order to fulfil the needs of the students and the duration of the semester.

Preface to the Second Edition

The first edition of the book has been used as the textbook for the standard graduate course in partial differential equations at York University since its publication in 1991. The motivation for writing the second edition stems from the desire to remove several deficiencies and obscurities, and to incorporate the improvements that I can see through many years of teaching the subject to graduate students and discussions of the subject with colleagues. Notwithstanding the many changes I have in mind, I am convinced that the elementary character of the book has served and will serve well as an ideal introduction to the study of pseudo-differential operators. Thus, the basic tenet of the second edition is to retain the style and the scope of the first edition.

Notable in the second edition is the addition of two chapters to the book. Experience in teaching pseudo-differential operators reveals the fact that many graduate students are still not comfortable with the interchange of order of integration and differentiation. The new chapter added to the beginning of the book is to prove a theorem to this effect which can cope with every interchange of order of integration and differentiation encountered in the book. Another new chapter, added as the final chapter in the second edition, is to prove a theorem on the existence of weak solutions of pseudo-differential equations. The inclusion of this chapter, in my opinion, enhances the value of the book as a book on partial differential equations. Furthermore, it provides a valuable connection with the chapter on minimal and maximal operators and the chapter on global regularity.

Other new features in the second edition include a deeper study of elliptic operators and parametrices, more details on the proof of the L^p-boundedness of pseudo-differential operators, additional exercises in several chapters of the book, a slightly expanded bibliography and an index.

Preface to the First Edition

The aim of the book is to give a straightforward account of a class of pseudo-differential operators. The prerequisite for understanding the book is a course in real variables. It is hoped that the book can be used in courses in functional analysis, Fourier analysis and partial differential equations.

The first eight chapters of the book contain the basic formal calculus of pseudo-differential operators. The remaining five chapters are devoted to some topics of a more functional analytic character.

It is clear to the expert that the book takes up a single theme in a wide subject and many important topics are omitted. It is my belief that this approach is in fact a more effective introduction of pseudo-differential operators to mathematicians and graduate students beginning to learn the subject. Exercises are included in the text. They are useful to anyone who wants to understand and appreciate the book better.

The actual writing of the book was essentially carried out and completed at the University of California at Irvine while I was on sabbatical leave from York University in the academic year in 1987–88. The preliminary drafts of the book have been used in seminars and graduate courses at the University of California at Irvine and York University.

Many colleagues and students have helped me improve the contents and organization of the book. In particular, I wish to thank Professor William Margulies at the California State University at Long Beach, Professor Martin Schechter at the University of California at Irvine, Professor Tuan Vu and Mr. Zhengbin Wang at York University for their stimulating conversations and critical comments about my book. I also wish to thank Mr. Lian Pi, my Ph.D. research student at York University, who has worked out every exercise in the book.

Contents

Chapter 1

Introduction, Notation and Preliminaries

Let $\mathbb{R}^n$ be the usual Euclidean space given by

$$\mathbb{R}^n = \{(x_1, x_2, \ldots, x_n) : x_j\text{'s are real numbers}\}.$$

We denote points in $\mathbb{R}^n$ by x, y, ξ, η etc. Let $x = (x_1, x_2, \ldots, x_n)$ and $y = (y_1, y_2, \ldots, y_n)$ be any two points in $\mathbb{R}^n$. The inner product $x \cdot y$ of x and y is defined by

$$x \cdot y = \sum_{j=1}^{n} x_j y_j$$

and the norm $|x|$ of x is defined by

$$|x| = \left(\sum_{j=1}^{n} x_j^2 \right)^{1/2}.$$

On $\mathbb{R}^n$, the simplest differential operators are $\frac{\partial}{\partial x_j}$, $j = 1, 2, \ldots, n$. We sometimes denote $\frac{\partial}{\partial x_j}$ by ∂_j. For reasons we shall see later in the book, we usually find the operator D_j given by $D_j = -i\partial_j$, $i^2 = -1$, better in expressing certain formulas.

The most general linear partial differential operator of order m on $\mathbb{R}^n$ treated in this book is of the form

$$\sum_{\alpha_1 + \alpha_2 + \cdots + \alpha_n \leq m} a_{\alpha_1, \alpha_2, \ldots, \alpha_n}(x) D_1^{\alpha_1} D_2^{\alpha_2} \cdots D_n^{\alpha_n}, \tag{1.1}$$

where $\alpha_1, \alpha_2, \ldots, \alpha_n$ are nonnegative integers and $a_{\alpha_1, \alpha_2, \ldots, \alpha_n}(x)$ is an infinitely differentiable complex-valued function on $\mathbb{R}^n$. To simplify the expression (1.1), we let

$$\alpha = (\alpha_1, \alpha_2, \ldots, \alpha_n),$$

1

$$|\alpha| = \sum_{j=1}^{n} \alpha_j$$

and

$$D^\alpha = D_1^{\alpha_1} D_2^{\alpha_2} \cdots D_n^{\alpha_n}.$$

The α, given by an n-tuple of nonnegative integers, is called a *multi-index*. We call $|x|$ the *length* of the multi-index α. With the help of multi-indices, we can rewrite our differential operator (1.1) in the better form

$$\sum_{|\alpha| \leq m} a_\alpha(x) D^\alpha. \tag{1.2}$$

For each fixed x in $\mathbb{R}^n$, the operator (1.2) is a polynomial in $D_1, D_2, \ldots, D_n$. Therefore it is natural to denote the operator (1.2) by $P(x, D)$. If we replace D in (1.2) by a point $\xi = (\xi_1, \xi_2, \ldots, \xi_n)$ in $\mathbb{R}^n$, then we obtain a polynomial $\sum_{|\alpha| \leq m} a_\alpha(x) \xi^\alpha$ in $\mathbb{R}^n$, where $\xi^\alpha = \xi_1^{\alpha_1} \xi_2^{\alpha_2} \cdots \xi_n^{\alpha_n}$. Naturally, this polynomial is denoted by $P(x, \xi)$. We call $P(x, \xi)$ the *symbol* of the operator $P(x, D)$.

In this book, we shall study the partial differential operators (1.2) and their generalizations called pseudo-differential operators. To do this, we find it convenient to introduce in Chapters 2–5 certain aspects of analysis pertinent to our need.

The following list of remarks, notation and formulas will be useful to us.

(i) We denote the set of all real numbers by $\mathbb{R}$ and the set of all complex numbers by $\mathbb{C}$.

(ii) All vector spaces are assumed to be over the field of complex numbers. All functions are assumed to be complex-valued unless otherwise specified.

(iii) We do not bother to distinguish a function f from its value $f(x)$ at x. In other words, we shall occasionally use the symbol $f(x)$ to denote the function f without any warning.

(iv) Although the differential operator $D^\alpha = D_1^{\alpha_1} D_2^{\alpha_2} \cdots D_n^{\alpha_n}$ is more useful to us, we still use the differential operator $\partial^\alpha = \partial_1^{\alpha_1} \partial_2^{\alpha_2} \cdots \partial_n^{\alpha_n}$ very often in the book. In case we want to emphasize the variable x (or ξ) with respect to which we differentiate, we write ∂_x^α (or ∂_ξ^α) for ∂^α and D_x^α (or D_ξ^α) for D^α.

(v) We denote the set of all infinitely differentiable functions on $\mathbb{R}^n$ by $C^\infty(\mathbb{R}^n)$.

(vi) The L^p norm of a function f in $L^p(\mathbb{R}^n)$, $1 \leq p \leq \infty$, is denoted by $\|f\|_p$.

Let $\alpha = (\alpha_1, \alpha_2, \ldots, \alpha_n)$ and $\beta = (\beta_1, \beta_2, \ldots, \beta_n)$ be any two multi-indices.

(vii) $\beta \leq \alpha$ means that $\beta_j \leq \alpha_j$ for $j = 1, 2, \ldots, n$.

(viii) $\alpha - \beta$ is the multi-index $(\alpha_1 - \beta_1, \alpha_2 - \beta_2, \ldots, \alpha_n - \beta_n)$ whenever $\beta \leq \alpha$.

(ix) $\alpha! = \alpha_1! \alpha_2! \cdots \alpha_n!$.

(x) $\binom{\alpha}{\beta} = \binom{\alpha_1}{\beta_1}\binom{\alpha_2}{\beta_2}\cdots\binom{\alpha_n}{\beta_n}$ whenever $\beta \leq \alpha$.

(xi) The formula

$$D^\alpha(fg) = \sum_{\beta \leq \alpha} \binom{\alpha}{\beta}(D^\beta f)(D^{\alpha-\beta}g) \tag{1.3}$$

is known as Leibniz's formula. It is a special case of the following more general Leibniz's formula.

(xii) Let $P(D) = \sum_{|\alpha|\leq m} a_\alpha D^\alpha$ be a linear partial differential operator with constant coefficients, and $P(\xi)$ its symbol. Then

$$P(D)(fg) = \sum_{|\mu|\leq m} \frac{1}{\mu!}(P^{(\mu)}(D)f)(D^\mu g),$$

where $P^{(\mu)}(D)$ is the linear partial differential operator with symbol $P^{(\mu)}(\xi)$ given by

$$P^{(\mu)}(\xi) = (\partial^\mu P)(\xi), \quad \xi \in \mathbb{R}^n.$$

(xiii) Let $f \in C^\infty(\mathbb{R}^n)$. Then

$$D^\alpha\left(\frac{1}{f}\right) = \sum C_{\alpha^{(1)}, \alpha^{(2)}, \ldots, \alpha^{(k)}} \frac{(\partial^{\alpha^{(1)}}f)(\partial^{\alpha^{(2)}}f)\cdots(\partial^{\alpha^{(k)}}f)}{f^{k+1}}, \tag{1.4}$$

where $C_{\alpha^{(1)}, \alpha^{(2)}, \ldots, \alpha^{(k)}}$'s are constants and the sum is taken over all possible multi-indices $\alpha^{(1)}, \alpha^{(2)}, \ldots, \alpha^{(k)}$, which form a partition of α. The formula (1.4) is valid at all points x in $\mathbb{R}^n$ for which $f(x) \neq 0$.

(xiv) Let f be a measurable function on $\mathbb{R}^n \times \mathbb{R}^n$. Then for $1 \leq p < \infty$,

$$\left\{\int_{\mathbb{R}^n}\left|\int_{\mathbb{R}^n} f(x,y)\,dy\right|^p dx\right\}^{1/p} \leq \int_{\mathbb{R}^n}\left\{\int_{\mathbb{R}^n}|f(x,y)|^p dx\right\}^{1/p} dy. \tag{1.5}$$

This inequality is the well-known *Minkowski's inequality in integral form*.

(xv) The inequality

$$|x^\alpha| \leq |x|^{|\alpha|} \tag{1.6}$$

for all $x \in \mathbb{R}^n$ and multi-indices α, which will be used quite often in this book, is an inequality in terms of the absolute value of a real number, the norm of a point in $\mathbb{R}^n$ and the length of a multi-index. Its proof is left as an exercise.

Exercises

1.1. Find the symbol of each of the following partial differential operators on $\mathbb{R}^2$.

(i) $\frac{\partial^2}{\partial x_1^2} + \frac{\partial^2}{\partial x_2^2}$

(ii) $\frac{\partial^2}{\partial x_1^2} - \frac{\partial^2}{\partial x_2^2}$

(iii) $\frac{\partial}{\partial x_1} + \frac{\partial^2}{\partial x_2^2}$

(iv) $\frac{\partial}{\partial x_1} + i\frac{\partial^2}{\partial x_2^2}$

(v) $\frac{\partial}{\partial x_1} + i\frac{\partial}{\partial x_2}$

1.2. For each of the partial differential operators in Exercise 1.1, find the zero set of the symbol, i.e., the set of zeros of the symbol.

1.3. Find the symbol of the partial differential operator

$$P(x, D) = \frac{\partial^2}{\partial x_1^2} + x_1^2 \frac{\partial^2}{\partial x_2^2}$$

on $\mathbb{R}^2$. For each fixed $x \in \mathbb{R}^2$, find the zero set $\{\xi \in \mathbb{R}^2 : P(x, \xi) = 0\}$.

1.4. What is the analog of Minkowski's inequality in (1.5) when $p = \infty$?

1.5. Prove inequality (1.6).

Chapter 2

Differentiation of Integrals Depending on Parameters

The aim of this chapter is to prove a theorem on how to differentiate an integral depending on parameters in order to justify every interchange of integration and differentiation throughout the book. I hope analysts can find the theorem, or some variant of it, useful. Other criteria can be found on, e.g., p.288 in [Friedman (1971)], pp.94–95 in [Royden (1988)], and p.85 in [Wheeden and Zygmund (1977)].

Theorem 2.1. *Let (Y, μ) be a measure space and $f : \mathbb{R}^n \times Y \to \mathbb{C}$ be a measurable function such that*
(i) $f(x, \cdot) \in L^1(Y)$ *for all x in $\mathbb{R}^n$,*
(ii) $f(\cdot, y) \in C^\infty(\mathbb{R}^n)$ *for almost all y in Y,*
(iii) $\sup_{x \in \mathbb{R}^n} \int_Y |(\partial_x^\alpha f)(x, y)| d\mu < \infty$ *for all multi-indices α.*
Then the integral $\int_Y f(x, y) d\mu$, as a function of x, is in $C^\infty(\mathbb{R}^n)$ and

$$\partial^\beta \int_Y f(x, y) \, d\mu = \int_Y (\partial_x^\beta f)(x, y) \, d\mu, \quad x \in \mathbb{R}^n,$$

for all multi-indices β.

We begin with a lemma.

Lemma 2.2. *Let $f : \mathbb{R}^n \times Y \to \mathbb{C}$ be such that the hypotheses of Theorem 2.1 are satisfied. Then for any multi-index α and $j = 1, 2, \ldots, n$, the integrals*

$$\int_Y (\partial_x^\alpha f)(x_1, \ldots, x_j, \ldots, x_n, y) \, d\mu$$

and

$$\int_Y |(\partial_x^\alpha f)(x_1, \ldots, x_j, \ldots, x_n, y)| \, d\mu,$$

as functions of x_j, are continuous on $\mathbb{R}$.

5

Proof Using the mean value theorem and hypothesis (iii), we obtain

$$\int_Y \{|(\partial_x^\alpha f)(x_1, \ldots, x_j + h, \ldots, x_n, y)| -$$

$$|(\partial_x^\alpha f)(x_1, \ldots, x_j, \ldots, x_n, y)|\} \, d\mu$$

$$\leq M_{\alpha,j}|h|$$

for all x in $\mathbb{R}^n$, where

$$M_{\alpha,j} = \sup_{x \in \mathbb{R}^n} \int_Y |(\partial_x^\alpha \partial_{x_j} f)(x, y)| \, d\mu,$$

and hence the lemma. □

Proof of Theorem 2.1 In view of of the proof of Lemma 2.2, the theorem is valid for the zero multi-index. Suppose that the theorem is valid for any multi-index with length l and let γ be a multi-index with length $l + 1$. If we write γ as $\beta + \varepsilon$, where β is a multi-index with length l, and ε is a multi-index with length one and the only nonzero entry in the jth position, then, by the fundamental theorem of calculus, Lemma 2.2 and the Fubini theorem,

$$\partial^\gamma \int_Y f(x, y) \, d\mu$$

$$= \partial_j \int_Y (\partial_x^\beta f)(x, y) \, d\mu$$

$$= \lim_{h \to 0} \frac{1}{h} \int_Y \{(\partial_x^\beta f)(x_1, \ldots, x_j + h, \ldots, x_n, y) -$$

$$(\partial_x^\beta f)(x_1, \ldots, x_j, \ldots, x_n, y)\} \, d\mu$$

$$= \lim_{h \to 0} \frac{1}{h} \int_Y \left\{ \int_{x_j}^{x_j + h} (\partial_x^\gamma f)(x_1, \ldots, s, \ldots, x_n, y) \, ds \right\} d\mu$$

$$= \lim_{h \to 0} \frac{1}{h} \int_{x_j}^{x_j + h} \left\{ \int_Y (\partial_x^\gamma f)(x_1, \ldots, s, \ldots, x_n, y) \, d\mu \right\} ds$$

$$= \int_Y (\partial_x^\gamma f)(x, y) \, d\mu, \quad x \in \mathbb{R}^n.$$

Thus, by induction, the proof is complete. □

From the proof of Theorem 2.1, we obtain the following result.

Corollary 2.3. *The conclusions of Theorem 2.1 remain valid if hypothesis* (iii) *is replaced by the hypothesis that, for every multi-index* α, *the integral* $\int_Y (\partial_x^\gamma f)(x, y) \, d\mu$, *as a function of* x, *is continuous on* $\mathbb{R}^n$.

Exercises

2.1. Let f be a bounded function defined on the strip

$$Q = \{(x_1, x_2) : x_1 \in \mathbb{R},\ 0 \le x_2 \le 1\}$$

in $\mathbb{R}^2$ such that for each fixed x_1 in $\mathbb{R}$, the function $f(x_1, \cdot)$ is measurable on $[0, 1]$. Furthermore, suppose that $(\partial_1^k f)(x_1, x_2)$ exists for all nonnegative integers k and all $(x_1, x_2) \in Q$, and for each nonnegative integer k, there exists a positive constant C_k such that

$$|(\partial_1^k f)(x_1, x_2)| \le C_k, \quad (x_1, x_2) \in Q.$$

Prove that

$$\left(\frac{d}{dx_1}\right)^k \int_0^1 f(x_1, x_2)\, dx_2 = \int_0^1 (\partial_1^k f)(x_1, x_2)\, dx_2$$

for all $x_1 \in \mathbb{R}$.

2.2. Let $f \in L^1(\mathbb{R}^n)$ and let $g \in C^\infty(\mathbb{R}^n)$ be such that $\partial^\alpha g \in L^\infty(\mathbb{R}^n)$ for all multi-indices α. Prove that the function h on $\mathbb{R}^n$ defined by

$$h(x) = \int_{\mathbb{R}^n} f(y)g(x - y)\, dy, \quad x \in \mathbb{R}^n,$$

is in $C^\infty(\mathbb{R}^n)$ and

$$(\partial^\alpha h)(x) = \int_{\mathbb{R}^n} f(y)(\partial^\alpha g)(x - y)\, dy, \quad x \in \mathbb{R}^n.$$

2.3. Let f be the function on $\mathbb{R}$ defined by

$$f(x) = \sum_{n=-\infty}^{\infty} c_n e^{inx}, \quad x \in \mathbb{R},$$

where $\ldots, c_{-2}, c_{-1}, c_0, c_1, c_2, \ldots$ are constants such that

$$\sum_{n=-\infty}^{\infty} |c_n| < \infty.$$

Prove that, if

$$\sum_{n=-\infty}^{\infty} |c_n| n^k < \infty$$

for all nonnegative integers k, then $f \in C^\infty(\mathbb{R})$ and for $k = 0, 1, 2, \ldots$, we have

$$f^{(k)}(x) = \sum_{n=-\infty}^{\infty} c_n n^k e^{inx}, \quad x \in \mathbb{R}.$$

Chapter 3

The Convolution

In this chapter we introduce two important subsets of $C^\infty(\mathbb{R}^n)$, usually denoted by $C_0^\infty(\mathbb{R}^n)$ and $\mathcal{S}$. The aim of this chapter is to prove that they are dense in $L^p(\mathbb{R}^n)$, $1 \leq p < \infty$. To this end, we need the notion of convolution.

Theorem 3.1. (Young's Inequality) *Let $f \in L^1(\mathbb{R}^n)$ and $g \in L^p(\mathbb{R}^n)$, $1 \leq p \leq \infty$. Then the integral*

$$\int_{\mathbb{R}^n} f(x-y)g(y)\,dy$$

exists for almost every $x \in \mathbb{R}^n$. If the value of the integral is denoted by $(f*g)(x)$, then $f*g \in L^p(\mathbb{R}^n)$ and

$$\|f*g\|_p \leq \|f\|_1\|g\|_p.$$

Remark 3.2. We usually call $f*g$ the *convolution* of f and g.

Proof of Theorem 3.1 For $p = 1$, let

$$h(x) = \int_{\mathbb{R}^n} |f(x-y)|\,|g(y)|\,dy.$$

Then, by Fubini's theorem,

$$
\begin{aligned}
\int_{\mathbb{R}^n} h(x)\,dx &= \int_{\mathbb{R}^n} \left(\int_{\mathbb{R}^n} |f(x-y)|\,|g(y)|\,dy \right) dx \\
&= \left(\int_{\mathbb{R}^n} |g(y)|\,dy \right) \left(\int_{\mathbb{R}^n} |f(x-y)|\,dx \right) \\
&= \|g\|_1\|f\|_1.
\end{aligned}
$$

Hence $h(x) < \infty$ for almost every $x \in \mathbb{R}^n$. Using Fubini's theorem again, we have

$$
\begin{aligned}
\|f * g\|_1 &= \int_{\mathbb{R}^n} \left| \int_{\mathbb{R}^n} f(x-y)g(y)\,dy \right| dx \\
&\leq \int_{\mathbb{R}^n} \left(\int_{\mathbb{R}^n} |f(x-y)|\,|g(y)|\,dy \right) dx \\
&= \left(\int_{\mathbb{R}^n} |g(y)|\,dy \right) \left(\int_{\mathbb{R}^n} |f(x-y)|\,dx \right) \\
&= \|g\|_1 \|f\|_1.
\end{aligned}
$$

This proves the theorem for $p = 1$. For $1 < p < \infty$, let

$$
h_p(x) = \int_{\mathbb{R}^n} |f(x-y)|\,|g(y)|^p dy.
$$

Then, as in the previous case, $h_p(x) < \infty$ for almost every $x \in \mathbb{R}^n$. Let p' be the conjugate index of p. Then, by Hölder's inequality,

$$
\begin{aligned}
&\int_{\mathbb{R}^n} |f(x-y)|\,|g(y)|\,dy \\
&= \int_{\mathbb{R}^n} |f(x-y)|^{1/p'} |f(x-y)|^{1/p} |g(y)|\,dy \\
&\leq \left\{ \int_{\mathbb{R}^n} |f(x-y)|\,dy \right\}^{1/p'} \left\{ \int_{\mathbb{R}^n} |f(x-y)|\,|g(y)|^p dy \right\}^{1/p} \\
&= \|f\|_1^{1/p'} \{h_p(x)\}^{1/p}.
\end{aligned} \tag{3.1}
$$

Hence $\int_{\mathbb{R}^n} f(x-y)g(y)dy$ exists for almost every $x \in \mathbb{R}^n$. Moreover, by (3.1), we have

$$
\begin{aligned}
\|f * g\|_p &= \left\{ \int_{\mathbb{R}^n} \left| \int_{\mathbb{R}^n} f(x-y)g(y)\,dy \right|^p dx \right\}^{1/p} \\
&\leq \left\{ \int_{\mathbb{R}^n} \left(\int_{\mathbb{R}^n} |f(x-y)|\,|g(y)|\,dy \right)^p dx \right\}^{1/p} \\
&\leq \|f\|_1^{1/p'} \left\{ \int_{\mathbb{R}^n} h_p(x)dx \right\}^{1/p} \\
&\leq \|f\|_1^{1/p'} \{\|f\|_1 \|G\|_1\}^{1/p},
\end{aligned}
$$

where $G(x) = |g(x)|^p$ for all $x \in \mathbb{R}^n$. Hence

$$
\|f * g\|_p \leq \|f\|_1 \|g\|_p.
$$

To prove the theorem for $p = \infty$, note that

$$\int_{\mathbb{R}^n} |f(x-y)|\,|g(y)|\,dy \leq \|g\|_\infty \int_{\mathbb{R}^n} |f(x-y)|\,dy = \|g\|_\infty \|f\|_1. \qquad (3.2)$$

Hence the integral $\int_{\mathbb{R}^n} f(x-y)\,g(y)\,dy$ exists for every $x \in \mathbb{R}^n$. Moreover, by (3.2), we have

$$\|f * g\|_\infty \leq \|g\|_\infty \|f\|_1.$$

$\square$

Proposition 3.3. (L^p-Continuity of Translations) *Let $f \in L^p(\mathbb{R}^n)$, $1 \leq p < \infty$. Then*

$$\lim_{|x| \to 0} \|f_x - f\|_p = 0,$$

where f_x is the function defined by

$$f_x(y) = f(x+y), \quad y \in \mathbb{R}^n.$$

Before proving Proposition 3.3, let us define $C_0(\mathbb{R}^n)$ to be the set of all continuous functions on $\mathbb{R}^n$ with compact support. The support of a continuous function h is defined to be the closure in $\mathbb{R}^n$ of the set

$$\{x \in \mathbb{R}^n : h(x) \neq 0\}$$

and is denoted by $\mathrm{supp}(h)$. We give one property of the set $C_0(\mathbb{R}^n)$ in the following proposition.

Proposition 3.4. *$C_0(\mathbb{R}^n)$ is dense in $L^p(\mathbb{R}^n)$ for $1 \leq p < \infty$.*

Proposition 3.4 is a measure-theoretic result which I ask you to believe. Otherwise, see p.69 in [Rudin (1987)] for a proof.

Proof of Proposition 3.3 Let $\delta > 0$ and $f \in L^p(\mathbb{R}^n)$. Then, by Proposition 3.4, there is a function g in $C_0(\mathbb{R}^n)$ such that

$$\|f - g\|_p < \frac{\delta}{3}. \qquad (3.3)$$

Now, using the triangle inequality and (3.3),

$$\|f_x - f\|_p \leq \|f_x - g_x\|_p + \|g_x - g\|_p + \|g - f\|_p < \frac{\delta}{3} + \frac{\delta}{3} + \frac{\delta}{3} = \delta$$

if $|x|$ is small enough. This completes the proof. $\square$

Theorem 3.5. *Let $\varphi \in L^1(\mathbb{R}^n)$ be such that*

$$\int_{\mathbb{R}^n} \varphi(x)\,dx = a.$$

For $\varepsilon > 0$, define the function φ_ε by

$$\varphi_\varepsilon(x) = \varepsilon^{-n}\varphi\left(\frac{x}{\varepsilon}\right), \quad x \in \mathbb{R}^n.$$

Then for any function $f \in L^p(\mathbb{R}^n), 1 \le p < \infty$, we have

$$f * \varphi_\varepsilon \to af$$

in $L^p(\mathbb{R}^n)$ as $\varepsilon \to 0$.

Proof Since

$$\int_{\mathbb{R}^n} \varphi_\varepsilon(x)\,dx = a$$

for all $\varepsilon > 0$, it follows from Minkowski's inequality in integral form given by (1.5) that we have

$$
\begin{aligned}
\|f * \varphi_\varepsilon - af\|_p &= \left(\int_{\mathbb{R}^n} |(f * \varphi_\varepsilon)(x) - af(x)|^p dx\right)^{1/p} \\
&= \left(\int_{\mathbb{R}^n}\left|\int_{\mathbb{R}^n}\{f(x-y)-f(x)\}\varphi_\varepsilon(y)\,dy\right|^p dx\right)^{1/p} \\
&= \left(\int_{\mathbb{R}^n}\left|\int_{\mathbb{R}^n}\{f(x-\varepsilon y)-f(x)\}\varphi(y)\,dy\right|^p dx\right)^{1/p} \\
&\le \int_{\mathbb{R}^n}\left\{\int_{\mathbb{R}^n}|f(x-\varepsilon y)-f(x)|^p|\varphi(y)|^p dx\right\}^{1/p} dy \\
&= \int_{\mathbb{R}^n}|\varphi(y)|\left\{\int_{\mathbb{R}^n}|f(x-\varepsilon y)-f(x)|^p dx\right\}^{1/p} dy \\
&= \int_{\mathbb{R}^n}|\varphi(y)|\,\|f_{-\varepsilon y}-f\|_p\,dy.
\end{aligned}
\tag{3.4}
$$

By Proposition 3.3, $\|f_{-\varepsilon y} - f\|_p \to 0$ as $\varepsilon \to 0$. Also, by the triangle inequality, $\|f_{-\varepsilon y} - f\|_p \le 2\|f\|_p$. Hence an application of the Lebsegue dominated convergence theorem to the last integral in (3.4) implies that

$$\|f * \varphi_\varepsilon - af\|_p \to 0$$

as $\varepsilon \to 0$. $\square$

We introduce two important function spaces. It is customary to denote by $C_0^\infty(\mathbb{R}^n)$ the set of all infinitely differentiable functions on $\mathbb{R}^n$ with compact support and by $\mathcal{S}$ the set of all infinitely differentiable functions φ on $\mathbb{R}^n$ such that for all multi-indices α and β,

$$\sup_{x \in \mathbb{R}^n} |x^\alpha(D^\beta\varphi)(x)| < \infty.$$

The space $\mathcal{S}$ is usually called the *Schwartz space* in deference to Laurent Schwartz. Obviously, $C_0^\infty(\mathbb{R}^n)$ is included in $\mathcal{S}$. That the inclusion is proper can be seen easily by noting that the function $e^{-|x|^2}$ is in $\mathcal{S}$ but not in $C_0^\infty(\mathbb{R}^n)$. We want to prove that $C_0^\infty(\mathbb{R}^n)$ is dense in $L^p(\mathbb{R}^n)$ for $1 \le p < \infty$. To this end, we need two preliminary results.

Proposition 3.6. *Let $\varphi \in \mathcal{S}$ and $f \in L^p(\mathbb{R}^n)$, $1 \le p \le \infty$. Then*

$$f * \varphi \in C^\infty(\mathbb{R}^n)$$

and

$$\partial^\alpha(f * \varphi) = f * (\partial^\alpha \varphi)$$

for every multi-index α.

Proof Let $\varphi \in \mathcal{S}$. Then for every multi-index α, $\partial^\alpha \varphi \in L^{p'}(\mathbb{R}^n)$, where p' is the conjugate index of p. Hence, by Hölder's inequality, $(f * (\partial^\alpha \varphi))(x)$ exists for every $x \in \mathbb{R}^n$. Therefore differentiation and integration can be interchanged. $\qquad\square$

Proposition 3.7. *Let f and g be continuous functions on $\mathbb{R}^n$ with compact support. Then the convolution $f * g$ also has compact support. In fact,*

$$\mathrm{supp}(f * g) \subseteq \mathrm{supp}(f) + \mathrm{supp}(g).$$

Remark 3.8. Let A and B be subsets of $\mathbb{R}^n$. Then the vector sum $A + B$ is defined by

$$A + B = \{x + y : x \in A \text{ and } y \in B\}.$$

In fact, the vector sum can be defined for any two subsets of a vector space.

Proof of Proposition 3.7 Since

$$(f * g)(x) = \int_{\mathbb{R}^n} f(x - y)g(y)\,dy,$$

it follows that if $(f * g)(x) \neq 0$, then there exists a $y \in \mathrm{supp}(g)$ such that $x - y \in \mathrm{supp}(f)$. Hence, by Remark 3.8,

$$x \in \mathrm{supp}(f) + \mathrm{supp}(g).$$

This completes the proof. $\qquad\square$

Theorem 3.9. $C_0^\infty(\mathbb{R}^n)$ *is dense in $L^p(\mathbb{R}^n)$ for $1 \le p < \infty$.*

Proof Let $\varphi \in C_0^\infty(\mathbb{R}^n)$ be such that

$$\int_{\mathbb{R}^n} \varphi(x)dx = 1.$$

Such a function exists by Exercise 3.2. For $\varepsilon > 0$, define φ_ε by

$$\varphi_\varepsilon(x) = \varepsilon^{-n}\varphi\left(\frac{x}{\varepsilon}\right), \quad x \in \mathbb{R}^n.$$

Then for all functions $g \in C_0(\mathbb{R}^n)$, we have, by Propositions 3.6 and 3.7, $g * \varphi_\varepsilon \in C_0^\infty(\mathbb{R}^n)$. Also, by Theorem 3.5,

$$g * \varphi_\varepsilon \to g \tag{3.5}$$

in $L^p(\mathbb{R}^n)$ as $\varepsilon \to 0$. Let $\delta > 0$ and $f \in L^p(\mathbb{R}^n)$. Then, by Proposition 3.4, there is a function $g \in C_0(\mathbb{R}^n)$ such that

$$\|f - g\|_p < \frac{\delta}{2}. \tag{3.6}$$

By (3.5), we can find a function $\psi \in C_0^\infty(\mathbb{R}^n)$ such that

$$\|g - \psi\|_p < \frac{\delta}{2}. \tag{3.7}$$

Hence, by the triangle inequality, (3.6) and (3.7), we have

$$\|f - \psi\|_p \leq \|f - g\|_p + \|g - \psi\|_p < \frac{\delta}{2} + \frac{\delta}{2} = \delta.$$

This proves that $C_0^\infty(\mathbb{R}^n)$ is dense in $L^p(\mathbb{R}^n)$ for $1 \leq p < \infty$. $\square$

Remark 3.10. An immediate consequence of Exercise 3.3 and Theorem 3.9 is that the Schwartz space $\mathcal{S}$ is also dense in $L^p(\mathbb{R}^n)$ for $1 \leq p < \infty$.

Exercises

3.1. Let φ and φ_ε be the functions given in the hypotheses of Theorem 3.5. Let f be a bounded function on $\mathbb{R}^n$ which is continuous on an open subset V of $\mathbb{R}^n$. Prove that $f * \varphi_\varepsilon \to af$ uniformly on every compact subset of V as $\varepsilon \to 0$.

3.2. Let φ be the function on $\mathbb{R}^n$ defined by

$$\varphi(x) = \begin{cases} \exp\left(-\frac{1}{1-|x|^2}\right), & |x| < 1, \\ 0, & |x| \geq 1. \end{cases}$$

Prove that $\varphi \in C_0^\infty(\mathbb{R}^n)$.

3.3. Prove that every function in S is in $L^p(\mathbb{R}^n)$, $1 \leq p \leq \infty$.

3.4. Is S dense in $L^\infty(\mathbb{R}^n)$? Explain your answer.

3.5. Determine whether or not the L^p-continuity of translations is true for $p = \infty$.

3.6. Use Minkowski's inequality in integral form in (1.5) to prove Young's inequality for $1 \leq p < \infty$.

Chapter 4

The Fourier Transform

The Fourier transform will be used in Chapter 6 to define pseudo-differential operators. Two important results in the theory of Fourier transforms are the Fourier inversion formula for Schwartz functions in $\mathcal{S}$ and the Plancherel theorem for functions in $L^2(\mathbb{R}^n)$. They are very useful for the study of pseudo-differential operators.

Let $f \in L^1(\mathbb{R}^n)$. We define $\hat{f}$ by

$$\hat{f}(\xi) = (2\pi)^{-n/2} \int_{\mathbb{R}^n} e^{-ix \cdot \xi} f(x)\, dx, \quad \xi \in \mathbb{R}^n.$$

The function $\hat{f}$ is called the *Fourier transform* of f and is sometimes denoted by $\mathcal{F}f$.

Proposition 4.1. *Let f and g be in $L^1(\mathbb{R}^n)$. Then*

$$\widehat{f * g} = (2\pi)^{n/2} \hat{f}\hat{g}.$$

Proof By Theorem 3.1, $f * g \in L^1(\mathbb{R}^n)$. Then, using the definition of the Fourier transform, we have

$$(2\pi)^{-n/2} \widehat{f * g}(\xi)$$
$$= (2\pi)^{-n} \int_{\mathbb{R}^n} e^{-ix \cdot \xi} (f * g)(x)\, dx$$
$$= (2\pi)^{-n} \int_{\mathbb{R}^n} e^{-ix \cdot \xi} \left(\int_{\mathbb{R}^n} f(x - y)\, g(y)\, dy \right) dx.$$

17

So, by Fubini's theorem, we get

$$(2\pi)^{-n/2}\widehat{f * g}(\xi)$$

$$= (2\pi)^{-n} \int_{\mathbb{R}^n} \int_{\mathbb{R}^n} e^{-i(x-y)\cdot\xi} f(x-y) e^{-iy\cdot\xi} g(y) \, dy \, dx$$

$$= (2\pi)^{-n} \int_{\mathbb{R}^n} e^{-iy\cdot\xi} g(y) \left(\int_{\mathbb{R}^n} e^{-i(x-y)\cdot\xi} f(x-y) \, dx \right) dy$$

$$= (2\pi)^{-n/2} \left(\int_{\mathbb{R}^n} e^{-iy\cdot\xi} g(y) \, dy \right) \hat{f}(\xi) = \hat{g}(\xi) \hat{f}(\xi).$$

$\square$

Proposition 4.2. *Let* $\varphi \in \mathcal{S}$. *Then*
(i) $(D^\alpha \varphi)^\wedge(\xi) = \xi^\alpha \hat{\varphi}(\xi)$ *for every multi-index* α,
(ii) $(D^\beta \hat{\varphi})(\xi) = ((-x)^\beta \varphi)^\wedge(\xi)$ *for every multi-index* β,
(iii) $\hat{\varphi} \in \mathcal{S}$.

Proof Integrating by parts, we get

$$(D^\alpha \varphi)^\wedge(\xi) = (2\pi)^{-n/2} \int_{\mathbb{R}^n} e^{-ix\cdot\xi} (D^\alpha \varphi)(x) \, dx$$

$$= (2\pi)^{-n/2} \int_{\mathbb{R}^n} \xi^\alpha e^{-ix\cdot\xi} \varphi(x) \, dx = \xi^\alpha \hat{\varphi}(\xi).$$

This proves part (i). For part (ii), we have

$$\left(D^\beta \hat{\varphi} \right)(\xi) = (2\pi)^{-n/2} D^\beta \left(\int_{\mathbb{R}^n} e^{-ix\cdot\xi} \varphi(x) \, dx \right)$$

$$= (2\pi)^{-n/2} \int_{\mathbb{R}^n} (-x)^\beta e^{-ix\cdot\xi} \varphi(x) \, dx = \left((-x)^\beta \varphi \right)^\wedge(\xi).$$

The interchange of the order of differentiation and integration is valid because $(-x)^\beta \varphi \in \mathcal{S}$. To prove part (iii), let α and β be any two multi-indices. Then, by parts (i) and (ii),

$$|\xi^\alpha (D^\beta \hat{\varphi})(\xi)| = |\xi^\alpha ((-x)^\beta \varphi)^\wedge(\xi)| = |\{D^\alpha ((-x)^\beta \varphi)\}^\wedge(\xi)|.$$

Since $D^\alpha ((-x)^\beta \varphi)$ is in $\mathcal{S}$, hence in $L^1(\mathbb{R}^n)$, it follows that

$$\sup_{\xi \in \mathbb{R}^n} |\xi^\alpha (D^\beta \hat{\varphi})(\xi)| = \sup_{\xi \in \mathbb{R}^n} |\{D^\alpha ((-x)^\beta \varphi)\}^\wedge(\xi)|$$

$$\leq (2\pi)^{-n/2} \|D^\alpha ((-x)^\beta \varphi)\|_1 < \infty.$$

$\square$

That the Fourier transform turns functions in $L^1(\mathbb{R}^n)$ into continuous functions vanishing at infinity is the content of the following proposition.

Proposition 4.3. (The Riemann–Lebesgue Lemma) *Let $f \in L^1(\mathbb{R}^n)$. Then*

(i) *$\hat{f}$ is continuous on $\mathbb{R}^n$,*

(ii) $\lim_{|\xi| \to \infty} \hat{f}(\xi) = 0$,

(iii) *$f_j \to f$ in $L^1(\mathbb{R}^n) \Rightarrow \hat{f}_j \to \hat{f}$ uniformly on $\mathbb{R}^n$.*

Proof Let $f_j \to f$ in $L^1(\mathbb{R}^n)$. Then

$$|\hat{f}_j(\xi) - \hat{f}(\xi)| \le (2\pi)^{-n/2} \|f_j - f\|_1.$$

Hence $\hat{f}_j \to \hat{f}$ uniformly on $\mathbb{R}^n$. This proves part (iii). To prove parts (i) and (ii), let $\varphi \in \mathcal{S}$. Then, by part (iii) of Proposition 4.2, $\hat{\varphi} \in \mathcal{S}$. Hence parts (i) and (ii) are satisfied for functions in $\mathcal{S}$. Let $f \in L^1(\mathbb{R}^n)$. Since $\mathcal{S}$ is dense in $L^1(\mathbb{R}^n)$, it follows that there is a sequence $\{\varphi_j\}$ of functions in $\mathcal{S}$ such that $\varphi_j \to f$ in $L^1(\mathbb{R}^n)$. By part (iii) which we have proved, $\widehat{\varphi_j} \to \hat{f}$ uniformly on $\mathbb{R}^n$. This proves parts (i) and (ii). $\qquad\square$

Let f be a measurable function defined on $\mathbb{R}^n$. For any fixed $y \in \mathbb{R}^n$, we define functions $T_y f$ and $M_y f$ by

$$(T_y f)(x) = f(x + y), \quad x \in \mathbb{R}^n, \tag{4.1}$$

and

$$(M_y f)(x) = e^{ix \cdot y} f(x), \quad x \in \mathbb{R}^n. \tag{4.2}$$

Let a be a nonzero real number. Then we define the function $D_a f$ by

$$(D_a f)(x) = f(ax), \quad x \in \mathbb{R}^n. \tag{4.3}$$

Proposition 4.4. *Let $f \in L^1(\mathbb{R}^n)$. Then the functions $T_y f, M_y f$ and $D_a f$ defined by (4.1), (4.2) and (4.3) respectively are in $L^1(\mathbb{R}^n)$. Moreover,*

(i) $(T_y f)^\wedge(\xi) = (M_y \hat{f})(\xi), \quad \xi \in \mathbb{R}^n$,

(ii) $(M_y f)^\wedge(\xi) = (T_{-y} \hat{f})(\xi), \quad \xi \in \mathbb{R}^n$,

(iii) $(D_a f)^\wedge(\xi) = |a|^{-n}(D_{1/a} \hat{f})(\xi), \quad \xi \in \mathbb{R}^n$.

Proof Obviously, $T_y f, M_y f$ and $D_a f$ are in $L^1(\mathbb{R}^n)$. By a simple change of variable, we have

$$
\begin{aligned}
(T_y f)^\wedge(\xi) &= (2\pi)^{-n/2} \int_{\mathbb{R}^n} e^{-ix\cdot\xi}(T_y f)(x)\, dx \\
&= (2\pi)^{-n/2} \int_{\mathbb{R}^n} e^{-ix\cdot\xi} f(x+y)\, dx \\
&= (2\pi)^{-n/2} \int_{\mathbb{R}^n} e^{-i(x-y)\cdot\xi} f(x)\, dx \\
&= e^{iy\cdot\xi}(2\pi)^{-n/2} \int_{\mathbb{R}^n} e^{-ix\cdot\xi} f(x)\, dx \\
&= e^{iy\cdot\xi}\hat{f}(\xi) \\
&= (M_y\hat{f})(\xi).
\end{aligned}
$$

Also,

$$
\begin{aligned}
(M_y f)^\wedge(\xi) &= (2\pi)^{-n/2} \int_{\mathbb{R}^n} e^{-ix\cdot\xi}(M_y f)(x)\, dx \\
&= (2\pi)^{-n/2} \int_{\mathbb{R}^n} e^{-ix\cdot\xi} e^{iy\cdot x} f(x)\, dx \\
&= \hat{f}(\xi - y) \\
&= (T_{-y}\hat{f})(\xi).
\end{aligned}
$$

Finally, by another change of variable, we have

$$
\begin{aligned}
(D_a f)^\wedge(\xi) &= (2\pi)^{-n/2} \int_{\mathbb{R}^n} e^{-ix\cdot\xi}(D_a f)(x)\, dx \\
&= (2\pi)^{-n/2} \int_{\mathbb{R}^n} e^{-ix\cdot\xi} f(ax)\, dx \\
&= (2\pi)^{-n/2} \int_{\mathbb{R}^n} e^{-i\left(\frac{x}{a}\right)\cdot\xi} f(x)|a|^{-n}\, dx \\
&= |a|^{-n}\hat{f}\left(\frac{\xi}{a}\right) \\
&= |a|^{-n}(D_{1/a}\hat{f})(\xi).
\end{aligned}
$$

$\square$

Proposition 4.5. *Let* $\varphi(x) = e^{-|x|^2/2}$. *Then* $\hat{\varphi}(\xi) = e^{-|\xi|^2/2}$.

Proof We first compute

$$
(2\pi)^{-n/2} \int_{\mathbb{R}^n} e^{-ix\cdot\xi-|x|^2}\, dx.
$$

Note that
$$(2\pi)^{-n/2} \int_{\mathbb{R}^n} e^{-ix\cdot\xi-|x|^2} dx = \prod_{j=1}^{n} (2\pi)^{-1/2} \int_{-\infty}^{\infty} e^{-ix_j\xi_j-x_j^2} dx_j. \qquad (4.4)$$

Hence it is sufficient to compute
$$(2\pi)^{-1/2} \int_{-\infty}^{\infty} e^{-it\zeta-t^2} dt, \quad \zeta \in (-\infty, \infty).$$

But
$$\int_{-\infty}^{\infty} e^{-it\zeta-t^2} dt = \int_{-\infty}^{\infty} e^{-(t^2+it\zeta)} dt$$

$$= e^{-\zeta^2/4} \int_{-\infty}^{\infty} e^{-(t^2+it\zeta-(\zeta^2/4))} dt$$

$$= e^{-\zeta^2/4} \int_{-\infty}^{\infty} e^{-(t+i(\zeta/2))^2} dt$$

$$= e^{-\zeta^2/4} \int_{L} e^{-z^2} dz, \qquad (4.5)$$

where L is the contour $\mathrm{Im}\, z = \frac{\zeta}{2}$ in the complex z-plane. Using Cauchy's integral theorem and the fact that the integrand goes to zero very fast as $|z| \to \infty$, we have
$$\int_{L} e^{-z^2} dz = \int_{-\infty}^{\infty} e^{-t^2} dt = \sqrt{\pi}. \qquad (4.6)$$

Hence, by (4.5) and (4.6),
$$(2\pi)^{-1/2} \int_{-\infty}^{\infty} e^{-it\zeta-t^2} dt = 2^{-1/2} e^{-\zeta^2/4}. \qquad (4.7)$$

By (4.4) and (4.7), we get
$$(2\pi)^{-n/2} \int_{\mathbb{R}^n} e^{-ix\cdot\xi-|x|^2} dx = 2^{-n/2} e^{-|\xi|^2/4}. \qquad (4.8)$$

Now, note that
$$(2\pi)^{-n/2} \int_{\mathbb{R}^n} e^{-ix\cdot\xi-(|x|^2/2)} dx = \left(D_{1/\sqrt{2}}\psi\right)^{\wedge}(\xi),$$

where $\psi(x) = e^{-|x|^2}$. Therefore, by (4.8) and part (iii) of Proposition 4.4, we get
$$\hat{\varphi}(\xi) = (2\pi)^{-n/2} \int_{\mathbb{R}^n} e^{-ix\cdot\xi-(|x|^2/2)} dx = e^{-|\xi|^2/2},$$

as asserted $\qquad\qquad\qquad\qquad\qquad\qquad\qquad\qquad\qquad\qquad\qquad\qquad \square$

Proposition 4.6. (The Adjoint Formula) *Let f and g be functions in $L^1(\mathbb{R}^n)$. Then*
$$\int_{\mathbb{R}^n} \hat{f}(x)g(x)\, dx = \int_{\mathbb{R}^n} f(x)\hat{g}(x)\, dx. \qquad (4.9)$$

Proof By Proposition 4.3, the Fourier transform of a function in $L^1(\mathbb{R}^n)$ is bounded on $\mathbb{R}^n$. Hence the integrals in (4.9) exist. Moreover,

$$
\begin{aligned}
\int_{\mathbb{R}^n} \hat{f}(x)g(x)\,dx &= (2\pi)^{-n/2} \int_{\mathbb{R}^n} \left(\int_{\mathbb{R}^n} e^{-ix\cdot y} f(y)dy \right) g(x)\,dx \\
&= (2\pi)^{-n/2} \int_{\mathbb{R}^n} f(y) \left(\int_{\mathbb{R}^n} e^{-ix\cdot y} g(x)\,dx \right) dy \\
&= \int_{\mathbb{R}^n} f(y)\hat{g}(y)\,dy.
\end{aligned}
$$

The interchange of the order of integration can obviously be justified by Fubini's theorem. $\square$

We are now prepared to prove the first important result in the theory of the Fourier transform.

Theorem 4.7. (The Fourier Inversion Formula) $(\hat{f})^{\vee} = f$ *for all functions* $f \in \mathcal{S}$. *Here, the operation* $\vee$ *is defined by*

$$
\check{g}(x) = (2\pi)^{-n/2} \int_{\mathbb{R}^n} e^{ix\cdot\xi} g(\xi)\,d\xi, \quad g \in \mathcal{S}.
$$

Remark 4.8. The function $\check{g}$ is usually called the *inverse Fourier transform* of g.

Proof of Theorem 4.7 We have

$$
(\hat{f})^{\vee}(x) = (2\pi)^{-n/2} \int_{\mathbb{R}^n} e^{ix\cdot\xi} \hat{f}(\xi)\,d\xi.
$$

Let $\varepsilon > 0$. Define

$$
I_{\varepsilon}(x) = (2\pi)^{-n/2} \int_{\mathbb{R}^n} e^{ix\cdot\xi - (\varepsilon^2|\xi|^2/2)} \hat{f}(\xi)\,d\xi. \tag{4.10}
$$

Let

$$
g(\xi) = e^{ix\cdot\xi - (\varepsilon^2|\xi|^2/2)} = (M_x D_\varepsilon \varphi)(\xi), \tag{4.11}
$$

where

$$
\varphi(\xi) = e^{-|\xi|^2/2}. \tag{4.12}
$$

Then, by Propositions 4.4 and 4.5,

$$
\hat{g}(\eta) = (T_{-x}\varepsilon^{-n} D_{1/\varepsilon}\hat{\varphi})(\eta) = \varepsilon^{-n} e^{-|\eta - x|^2/(2\varepsilon^2)}. \tag{4.13}
$$

Hence, by (4.10), (4.11), (4.13) and Proposition 4.6,

$$
\begin{aligned}
I_\varepsilon(x) &= (2\pi)^{-n/2} \int_{\mathbb{R}^n} g(\xi)\hat{f}(\xi)\,d\xi \\
&= (2\pi)^{-n/2} \int_{\mathbb{R}^n} \hat{g}(\eta)f(\eta)\,d\eta \\
&= \varepsilon^{-n}(2\pi)^{-n/2} \int_{\mathbb{R}^n} e^{-|\eta-x|^2/(2\varepsilon^2)} f(\eta)\,d\eta \\
&= (2\pi)^{-n/2}(f * \varphi_\varepsilon)(x),
\end{aligned}
\tag{4.14}
$$

where $\varphi_\varepsilon(x) = \varepsilon^{-n}\varphi\left(\frac{x}{\varepsilon}\right)$. Since $f \in \mathcal{S}$, it follows that f is in $L^p(\mathbb{R}^n)$, $1 \le p < \infty$. Therefore, by (4.12), (4.14) and Theorem 3.5,

$$
I_\varepsilon \to (2\pi)^{-n/2}\left(\int_{\mathbb{R}^n} e^{-|x|^2/2}dx\right)f = f
$$

in $L^p(\mathbb{R}^n)$ as $\varepsilon \to 0$. Hence there exists a sequence $\{\varepsilon_n\}$ of positive real numbers such that $I_{\varepsilon_n}(x) \to f(x)$ for almost every $x \in \mathbb{R}^n$ as $\varepsilon_n \to 0$. By (4.10) and Lebesgue's dominated convergence theorem,

$$
I_\varepsilon(x) \to (2\pi)^{-n/2} \int_{\mathbb{R}^n} e^{ix\cdot\xi}\hat{f}(\xi)\,d\xi
$$

for every $x \in \mathbb{R}^n$ as $\varepsilon \to 0$. Hence

$$
(2\pi)^{-n/2} \int_{\mathbb{R}^n} e^{ix\cdot\xi}\hat{f}(\xi)\,d\xi = f(x)
$$

for every $x \in \mathbb{R}^n$, and the proof is complete. $\qquad\square$

Remark 4.9. An immediate consequence of the Fourier inversion formula is that the Fourier transformation $f \to \hat{f}$ is a one to one mapping of $\mathcal{S}$ onto $\mathcal{S}$. If we define $\tilde{f}$ by

$$
\tilde{f}(x) = f(-x), \quad x \in \mathbb{R}^n,
$$

then the Fourier inversion formula is equivalent to the formula

$$
\hat{\hat{f}} = \tilde{f}, \quad f \in \mathcal{S}.
$$

The next important result is the Plancherel theorem.

Theorem 4.10. (The Plancherel Theorem) *The mapping $f \mapsto \hat{f}$ defined on $\mathcal{S}$ can be extended uniquely to a unitary operator on $L^2(\mathbb{R}^n)$.*

Proof Using the fact that $\mathcal{S}$ is dense in $L^2(\mathbb{R}^n)$ and the Fourier inversion formula, it is sufficient to prove that

$$\|\hat{\varphi}\|_2 = \|\varphi\|_2, \quad \varphi \in \mathcal{S}.$$

Let ψ be the function defined by

$$\psi(x) = \overline{\varphi(-x)}, \quad x \in \mathbb{R}^n. \tag{4.15}$$

Then $\psi \in \mathcal{S}$ and

$$\begin{aligned}
\hat{\psi}(\xi) &= (2\pi)^{-n/2} \int_{\mathbb{R}^n} e^{-ix\cdot\xi} \overline{\varphi(-x)} \, dx \\
&= (2\pi)^{-n/2} \int_{\mathbb{R}^n} e^{ix\cdot\xi} \overline{\varphi(x)} \, dx \\
&= \overline{\hat{\varphi}(\xi)}.
\end{aligned} \tag{4.16}$$

Thus, by (4.15), (4.16), the adjoint formula and the Fourier inversion formula,

$$\begin{aligned}
\|\hat{\varphi}\|_2^2 &= \int_{\mathbb{R}^n} \hat{\varphi}(\xi)\overline{\hat{\varphi}(\xi)} \, d\xi \\
&= \int_{\mathbb{R}^n} \hat{\varphi}(\xi)\hat{\psi}(\xi) \, d\xi = \int_{\mathbb{R}^n} \hat{\hat{\varphi}}(\xi)\psi(\xi) \, d\xi \\
&= \int_{\mathbb{R}^n} \tilde{\varphi}(\xi)\psi(\xi) \, d\xi = \int_{\mathbb{R}^n} \varphi(-\xi)\overline{\varphi(-\xi)} \, d\xi \\
&= \int_{\mathbb{R}^n} \varphi(\xi)\overline{\varphi(\xi)} \, d\xi = \|\varphi\|_2^2
\end{aligned}$$

and this completes the proof. $\qquad\square$

Remark 4.11. The Plancherel theorem states that the Fourier transform of a function in $L^2(\mathbb{R}^n)$ can be defined. If $f \in L^2(\mathbb{R}^n)$, then we shall denote its Fourier transform by $\hat{f}$ or $\mathcal{F}f$. The inverse of $\mathcal{F} : L^2(\mathbb{R}^n) \to L^2(\mathbb{R}^n)$ is of course denoted by $\mathcal{F}^{-1} : L^2(\mathbb{R}^n) \to L^2(\mathbb{R}^n)$.

Exercises

4.1. Here is another elegant proof of Proposition 4.5.
(i) Let φ be the function defined on $\mathbb{R}$ by

$$\varphi(x) = e^{-x^2/2}, \quad x \in \mathbb{R}.$$

Let $y = \hat{\varphi}$. Prove that

$$y'(\xi) + \xi y(\xi) = 0, \quad \xi \in \mathbb{R}.$$

(ii) Use the result in part (i) to prove that

$$\hat{\varphi}(\xi) = e^{-\xi^2/2}, \quad \xi \in \mathbb{R}.$$

4.2. Let $\{\varphi_n\}$ be the sequence of functions defined on $\mathbb{R}$ by

$$\varphi_0(x) = e^{-x^2/2}$$

and

$$\varphi_{n+1}(x) = x\varphi_n(x) - \varphi_n'(x)$$

for all $x \in \mathbb{R}$ and $n = 0, 1, 2, \ldots$. We call φ_n the *Hermite function of order n.*

(i) Prove that

$$\varphi_n(x) = (-1)^n e^{x^2/2} \left(\frac{d}{dx}\right)^n (e^{-x^2})$$

for all $x \in \mathbb{R}$ and $n = 0, 1, 2, \ldots$.

(ii) Prove that $\widehat{\varphi_n} = i^{-n}\varphi_n$ for $n = 0, 1, 2, \ldots$.

4.3. Prove that if $f \in L^1(\mathbb{R}^n)$ and $\hat{f} \in L^1(\mathbb{R}^n)$, then $(\hat{f})^\vee = f$ a.e.

4.4. Find a function f in $L^1(\mathbb{R}^n)$ such that $\hat{f}$ is not in $L^1(\mathbb{R}^n)$.

4.5. Prove that for all functions f in $L^1(\mathbb{R}^n)$,

$$\hat{f}(\xi) = \check{f}(-\xi), \quad \xi \in \mathbb{R}^n.$$

4.6. Prove that for all functions f in $L^1(\mathbb{R}^n)$,

$$\hat{\tilde{f}} = \tilde{\hat{f}}.$$

4.7. Prove that $\mathcal{F} : \mathcal{S} \to \mathcal{S}$ is bijective.

Chapter 5

Tempered Distributions

We only give the rudiments of the theory of tempered distributions in this chapter. More details on this subject will be introduced in later chapters as the need arises.

Definition 5.1. A sequence $\{\varphi_j\}$ of functions in the Schwartz space $\mathcal{S}$ is said to *converge to zero in* $\mathcal{S}$ (denoted by $\varphi_j \to 0$ in $\mathcal{S}$) if for all multi-indices α and β, we have

$$\sup_{x \in \mathbb{R}^n} |x^\alpha (D^\beta \varphi_j)(x)| \to 0$$

as $j \to \infty$.

Definition 5.2. A linear functional T on $\mathcal{S}$ is called a *tempered distribution* if for any sequence $\{\varphi_j\}$ of functions in $\mathcal{S}$ converging to zero in $\mathcal{S}$, we have

$$T(\varphi_j) \to 0$$

as $j \to \infty$.

Definition 5.3. Let f be a measurable function defined on $\mathbb{R}^n$ such that

$$\int_{\mathbb{R}^n} \frac{|f(x)|}{(1 + |x|)^N} dx < \infty$$

for some positive integer N. Then we call f a *tempered function*.

Proposition 5.4. *Let f be a tempered function defined on $\mathbb{R}^n$. Then the linear functional T_f on $\mathcal{S}$ defined by*

$$T_f(\varphi) = \int_{\mathbb{R}^n} f(x)\varphi(x) \, dx, \quad \varphi \in \mathcal{S},$$

is a tempered distribution.

Proof Let N be a positive integer such that

$$\int_{\mathbb{R}^n} \frac{|f(x)|}{(1+|x|)^N} dx < \infty.$$

Then for all functions $\varphi \in \mathcal{S}$, the integral $\int_{\mathbb{R}^n} f(x)\varphi(x)\, dx$ exists. Indeed, we have

$$\int_{\mathbb{R}^n} |f(x)|\,|\varphi(x)|\, dx$$

$$= \int_{\mathbb{R}^n} \frac{|f(x)|}{(1+|x|)^N}(1+|x|)^N|\varphi(x)|\, dx$$

$$\leq \left(\int_{\mathbb{R}^n} \frac{|f(x)|}{(1+|x|)^N} dx \right) \sup_{x \in \mathbb{R}^n} \{(1+|x|)^N|\varphi(x)|\} < \infty.$$

Let $\{\varphi_j\}$ be a sequence of functions in $\mathcal{S}$ converging to zero in $\mathcal{S}$ as $j \to \infty$. Then obviously,

$$\sup_{x \in \mathbb{R}^n} \{(1+|x|)^N|\varphi_j(x)|\} \to 0. \tag{5.1}$$

Since

$$|T_f(\varphi_j)| \leq \int_{\mathbb{R}^n} |f(x)|\,|\varphi_j(x)|\, dx$$

$$\leq \sup_{x \in \mathbb{R}^n} \{(1+|x|)^N|\varphi_j(x)|\} \int_{\mathbb{R}^n} \frac{|f(x)|}{(1+|x|)^N} dx \tag{5.2}$$

for all j, it follows from (5.1) and (5.2) that $T_f(\varphi_j) \to 0$ as $j \to \infty$. $\quad\square$

Proposition 5.5. *Let* $f \in L^p(\mathbb{R}^n)$, $1 \leq p \leq \infty$. *Then the linear functional* T_f *on* $\mathcal{S}$ *defined by*

$$T_f(\varphi) = \int_{\mathbb{R}^n} f(x)\varphi(x)\, dx, \quad \varphi \in \mathcal{S},$$

is a tempered distribution.

Proof Note that f is a tempered function. See Exercise 5.1. $\quad\square$

Remark 5.6. It is customary to identify the tempered distribution T_f with the function f and to say that such tempered distributions are functions.

Definition 5.7. Let T be a tempered distribution. Then the Fourier transform of T is defined to be the linear functional $\hat{T}$ on $\mathcal{S}$ given by

$$\hat{T}(\varphi) = T(\hat{\varphi}), \quad \varphi \in \mathcal{S}.$$

Proposition 5.8. $\hat{T}$ *is also a tempered distribution.*

Proof Let $\{\varphi_j\}$ be a sequence of functions in $\mathcal{S}$ converging to zero in $\mathcal{S}$. We need only prove that the sequence $\{\hat{\varphi}_j\}$ also converges to zero in $\mathcal{S}$. To this end, let α and β be any two multi-indices. Then, by Proposition 4.2,

$$\sup_{\xi \in \mathbb{R}^n} |\xi^\alpha (D^\beta \widehat{\varphi_j})(\xi)|$$

$$= \sup_{\xi \in \mathbb{R}^n} |\xi^\alpha ((-x)^\beta \varphi_j)^\wedge (\xi)|$$

$$= \sup_{\xi \in \mathbb{R}^n} |\{D^\alpha ((-x)^\beta \varphi_j)\}^\wedge (\xi)|$$

$$\leq (2\pi)^{-n/2} \|D^\alpha ((-x)^\beta \varphi_j)\|_1. \tag{5.3}$$

Since $\varphi_j \to 0$ in $\mathcal{S}$ as $j \to \infty$, it follows that for any positive integer N, we have

$$\sup_{x \in \mathbb{R}^n} \{(1 + |x|)^N |(D^\alpha ((-x)^\beta \varphi_j))(x)|\} \to 0 \tag{5.4}$$

as $j \to \infty$. Since for any positive integer N greater than n, we have

$$\|(D^\alpha ((-x)^\beta \varphi_j)\|_1$$

$$\leq \sup_{x \in \mathbb{R}^n} \{(1 + |x|)^N |(D^\alpha ((-x)^\beta \varphi_j))(x)|\} \int_{\mathbb{R}^n} (1 + |x|)^{-N} dx \tag{5.5}$$

for all j. Hence, by (4.3), (4.4) and (4.5), we conclude that $\hat{\varphi}_j \to 0$ in $\mathcal{S}$ as $j \to \infty$. $\qquad\square$

Theorem 5.9. (The Fourier Inversion Formula) *Let T be a tempered distribution. Then*

$$\hat{\hat{T}} = \tilde{T},$$

where $\tilde{T}$ is defined by

$$\tilde{T}(\varphi) = T(\tilde{\varphi}), \quad \varphi \in \mathcal{S}.$$

Proof Let $\varphi \in \mathcal{S}$. Then, by Definition 5.7 and the Fourier inversion formula for $\mathcal{S}$, we have

$$\hat{\hat{T}}(\varphi) = \hat{T}(\hat{\varphi}) = T(\hat{\hat{\varphi}}) = T(\tilde{\varphi}) = \tilde{T}(\varphi).$$

$\qquad\square$

Exercises

5.1. Prove that any function in $L^p(\mathbb{R}^n)$, $1 \le p \le \infty$, is a tempered function.

5.2. Let $\delta : \mathcal{S} \to \mathbb{C}$ be the mapping defined by

$$\delta(\varphi) = \varphi(0), \quad \varphi \in \mathcal{S}.$$

(i) Prove that δ is a tempered distribution.
(ii) Prove that δ is not a tempered function. (See Remark 5.6.)

5.3. Let $f \in L^1(\mathbb{R}^n)$ and T be the tempered distribution which is equal to f. Prove that $\hat{T}$ is equal to $\hat{f}$.

5.4. Do Exercise 5.3 again for $f \in L^2(\mathbb{R}^n)$.

5.5. Find the Fourier transform of the tempered distribution δ defined in Exercise 5.2.

5.6. Prove that there exist positive constants C_1 and C_2 such that

$$C_1(1 + |\xi|) \le (1 + |\xi|^2)^{1/2} \le C_2(1 + |\xi|), \quad \xi \in \mathbb{R}^n.$$

Chapter 6

Symbols, Pseudo-Differential Operators and Asymptotic Expansions

In this chapter we give the definition and the most elementary properties of a pseudo-differential operator and its symbol.

We begin by recalling that a linear partial differential operator $P(x, D)$ on $\mathbb{R}^n$ is given by

$$P(x, D) = \sum_{|\alpha| \leq m} a_\alpha(x) D^\alpha, \tag{6.1}$$

where the coefficients $a_\alpha(x)$ are functions defined on $\mathbb{R}^n$. If we replace the D^α in (6.1) by the monomial ξ^α in $\mathbb{R}^n$, then we obtain the so-called *symbol*

$$P(x, \xi) = \sum_{|\alpha| \leq m} a_\alpha(x) \xi^\alpha \tag{6.2}$$

of the operator (6.1). In order to get another representation of the operator $P(x, D)$, let us take any function φ in $\mathcal{S}$. Then, by (6.1), (6.2), Proposition 4.2 and the Fourier inversion formula for Schwartz functions, we have

$$
\begin{aligned}
(P(x, D)\varphi)(x) &= \sum_{|\alpha| \leq m} a_\alpha(x)(D^\alpha \varphi)(x) \\
&= \sum_{|\alpha| \leq m} a_\alpha(x)(\widehat{D^\alpha \varphi})^\vee(x) \\
&= \sum_{|\alpha| \leq m} a_\alpha(x)(\xi^\alpha \hat{\varphi})^\vee(x) \\
&= \sum_{|\alpha| \leq m} a_\alpha(x)(2\pi)^{-n/2} \int_{\mathbb{R}^n} e^{ix \cdot \xi} \xi^\alpha \hat{\varphi}(\xi) \, d\xi \\
&= (2\pi)^{-n/2} \int_{\mathbb{R}^n} e^{ix \cdot \xi} \left(\sum_{|\alpha| \leq m} a_\alpha(x) \xi^\alpha \right) \hat{\varphi}(\xi) \, d\xi \\
&= (2\pi)^{-n/2} \int_{\mathbb{R}^n} e^{ix \cdot \xi} P(x, \xi) \hat{\varphi}(\xi) \, d\xi.
\end{aligned}
\tag{6.3}
$$

So, we have represented the partial differential operator $P(x, D)$ in terms of its symbol by means of the Fourier transform. This representation immediately suggests that we can get operators more general than partial differential operators if we replace the symbol $P(x, \xi)$ by more general symbols $\sigma(x, \xi)$, which are no longer polynomials in ξ. The operators so obtained are called pseudo-differential operators. We shall do this in due course. Meanwhile, we should point out that in order to get a useful and tractable class of operators, it is necessary to impose certain conditions on the functions $\sigma(x, \xi)$. Many different sets of conditions have been proposed, resulting in many different classes of pseudo-differential operators. Our discussion in this book is first restricted to the following class.

Definition 6.1. Let $m \in (-\infty, \infty)$. Then we define S^m to be the set of all functions $\sigma(x, \xi)$ in $C^\infty (\mathbb{R}^n \times \mathbb{R}^n)$ such that for any two multi-indices α and β, there is a positive constant $C_{\alpha,\beta}$, depending on α and β only, for which

$$|(D_x^\alpha D_\xi^\beta \sigma)(x, \xi)| \leq C_{\alpha,\beta}(1 + |\xi|)^{m-|\beta|}, \quad x, \xi \in \mathbb{R}^n.$$

We call any function σ in $\cup_{m \in \mathbb{R}} S^m$ a *symbol*.

Definition 6.2. Let σ be a symbol. Then the *pseudo-differential operator* T_σ associated to σ is defined by

$$(T_\sigma \varphi)(x) = (2\pi)^{-n/2} \int_{\mathbb{R}^n} e^{ix \cdot \xi} \sigma(x, \xi) \hat{\varphi}(\xi) \, d\xi, \quad \varphi \in \mathcal{S}. \tag{6.4}$$

We give some examples.

Example 6.3. Let $P(x, D) = \sum_{|\alpha| \leq m} a_\alpha(x) D^\alpha$ be a linear partial differential operator on $\mathbb{R}^n$. If all the coefficients $a_\alpha(x)$ are C^∞ and have bounded derivatives of all orders, then the polynomial

$$P(x, \xi) = \sum_{|\alpha| \leq m} a_\alpha(x) \xi^\alpha$$

is in S^m and hence $P(x, D)$ is a pseudo-differential operator. (See formulas (6.3) and (6.4).)

Proof Let γ and δ be multi-indices. Then

$$|(D_x^\gamma D_\xi^\delta P)(x, \xi)| \leq \sum_{|\alpha| \leq m} C_{\alpha,\gamma} |\partial_\xi^\delta \xi^\alpha| \tag{6.5}$$

for all $x, \xi \in \mathbb{R}^n$, where

$$C_{\alpha,\gamma} = \sup_{x \in \mathbb{R}^n} |(D^\gamma a_\alpha)(x)|.$$

It can be shown easily that

$$\partial_\xi^\delta \xi^\alpha = \begin{cases} \delta! \binom{\alpha}{\delta} \xi^{\alpha-\delta}, & \delta \leq \alpha, \\ 0, & \text{otherwise}, \end{cases} \tag{6.6}$$

for all $\xi \in \mathbb{R}^n$. The proof of (6.6) is left as an exercise. See Exercise 6.5. Hence, by (1.6), (6.5) and (6.6),

$$|(D_x^\gamma D_\xi^\delta P)(x,\xi)| \leq \sum_{|\alpha| \leq m} C_{\alpha,\gamma} \delta! \binom{\alpha}{\delta} |\xi|^{|\alpha|-|\delta|}$$

$$\leq C'_{\gamma,\delta}(1 + |\xi|)^{m-|\delta|}$$

for all $x, \xi \in \mathbb{R}^n$, where

$$C'_{\gamma,\delta} = \sum_{|\alpha| \leq m} C_{\alpha,\gamma} \delta! \binom{\alpha}{\delta}.$$

$\square$

Example 6.4. Let $\sigma(\xi) = (1+|\xi|^2)^{m/2}$, $-\infty < m < \infty$. Then $\sigma \in S^m$ and hence T_σ is a pseudo-differential operator. It is sometimes advantageous to denote T_σ by $(I - \Delta)^{m/2}$, where I is the identity operator and Δ the Laplacian, i.e.,

$$\Delta = \sum_{j=1}^n \frac{\partial^2}{\partial x_j^2}.$$

Proof We need only prove that for any $m \in (-\infty, \infty)$ and multi-index β, there exists a positive constant $C_{m,\beta}$ such that

$$|(D^\beta \sigma)(\xi)| \leq C_{m,\beta}(1 + |\xi|)^{m-|\beta|} \tag{6.7}$$

for all $\xi \in \mathbb{R}^n$. (6.7) is obviously true for the zero multi-index. Suppose that (6.7) is true for all $m \in (-\infty, \infty)$ and multi-indices β of length at most equal to l. Let γ be a multi-index of length $l + 1$. Then $D^\gamma = D^\beta D_j$ for some $j = 1, 2, \ldots, n$ and some multi-index β of length l. Hence

$$|(D^\gamma \sigma)(\xi)| = |(\partial^\beta \partial_j \sigma)(\xi)| = |(\partial^\beta \tau)(\xi)| \tag{6.8}$$

for all $\xi \in \mathbb{R}^n$, where

$$\tau(\xi) = m\xi_j(1 + |\xi|^2)^{(m/2)-1}$$

for all $\xi \in \mathbb{R}^n$. Therefore, by Leibniz's formula,

$$(\partial^\beta \tau)(\xi) = m \sum_{\delta \leq \beta} \binom{\beta}{\delta} (\partial^\delta \xi_j) \partial^{\beta-\delta} \{(1 + |\xi|^2)^{(m/2)-1}\} \tag{6.9}$$

for all $\xi \in \mathbb{R}^n$. Hence, by (6.9) and the induction hypothesis, there exists a positive constant $C_{m,\beta}$ such that

$$|(\partial^\beta \tau)(\xi)| \leq C_{m,\beta} \sum_{\delta \leq \beta} \binom{\beta}{\delta}(1+|\xi|)^{1-|\delta|}(1+|\xi|)^{m-2-|\beta|+|\delta|}$$

$$= C'_{m,\beta}(1+|\xi|)^{m-|\gamma|}$$

for all $\xi \in \mathbb{R}^n$, where

$$C'_{m,\beta} = C_{m,\beta} \sum_{\delta \leq \beta} \binom{\beta}{\delta}.$$

Thus, by (6.8) and the principle of mathematical induction, the proof is complete. $\square$

We give two very simple properties of pseudo-differential operators.

Proposition 6.5. *Let σ and τ be two symbols such that $T_\sigma = T_\tau$. Then $\sigma = \tau$.*

We first prove a lemma.

Lemma 6.6. *Let f be a continuous tempered function such that*

$$T_f(\varphi) = 0, \quad \varphi \in \mathcal{S}. \tag{6.10}$$

Then f is identically zero on $\mathbb{R}^n$.

Proof Without loss of generality, we assume that f is real-valued. Suppose $f(x_0) \neq 0$ for some $x_0 \in \mathbb{R}^n$. Then there is an open ball $B(x_0, r)$ with center x_0 and radius r on which f is strictly positive (or negative). Choose a nonzero function $\varphi_0 \in C_0^\infty(\mathbb{R}^n)$ such that $\varphi_0(x) \geq 0$ for all $x \in \mathbb{R}^n$ and $\mathrm{supp}(\varphi_0) \subset B(x_0, r)$. Such a function exists by Exercise 3.2. It is clear that $\varphi_0 \in \mathcal{S}$ and

$$T_f(\varphi_0) = \int_{\mathbb{R}^n} f(x)\varphi_0(x)\,dx$$

is strictly positive (or negative). This contradicts (6.10). $\square$

Proof of Proposition 6.5 By hypothesis and Definition 6.2, we have

$$\int_{\mathbb{R}^n} e^{ix\cdot\xi}\{\sigma(x,\xi) - \tau(x,\xi)\}\hat{\varphi}(\xi)\,d\xi = 0, \quad \varphi \in \mathcal{S}.$$

Since $\mathcal{F} : \mathcal{S} \to \mathcal{S}$ is one to one and onto by the Fourier inversion formula, it follows that

$$\int_{\mathbb{R}^n} e^{ix\cdot\xi}\{\sigma(x,\xi) - \tau(x,\xi)\}\varphi(\xi)\,d\xi = 0, \quad \varphi \in \mathcal{S}.$$

Now, for any fixed $x \in \mathbb{R}^n$, $e^{ix\cdot\xi}\{\sigma(x,\xi) - \tau(x,\xi)\}$ is a continuous tempered function in the variable ξ. This follows easily from the definition of a symbol in Definition 6.1. Hence, by Lemma 6.6,

$$e^{ix\cdot\xi}\{\sigma(x,\xi) - \tau(x,\xi)\} = 0$$

for all $\xi \in \mathbb{R}^n$. Since x is fixed but arbitrary, it follows that

$$\sigma(x,\xi) - \tau(x,\xi) = 0$$

for all $x, \xi \in \mathbb{R}^n$. This proves that $\sigma = \tau$. $\qquad\square$

Proposition 6.7. *Let σ be a symbol. Then T_σ maps the Schwartz space $\mathcal{S}$ into itself.*

Proof Let $\varphi \in \mathcal{S}$. Then for any two multi-indices α and β, we need only prove that

$$\sup_{x\in\mathbb{R}^n} |x^\alpha (D^\beta (T_\sigma\varphi))(x)| < \infty.$$

But, using integration by parts and Leibniz's formula, we have

$$x^\alpha (D^\beta (T_\sigma\varphi))(x)$$

$$= x^\alpha (2\pi)^{-n/2} \int_{\mathbb{R}^n} D_x^\beta \{e^{ix\cdot\xi}\sigma(x,\xi)\}\hat{\varphi}(\xi)\,d\xi$$

$$= x^\alpha (2\pi)^{-n/2} \int_{\mathbb{R}^n} \sum_{\gamma\leq\beta} \binom{\beta}{\gamma} \xi^\gamma e^{ix\cdot\xi}(D_x^{\beta-\gamma}\sigma)(x,\xi)\hat{\varphi}(\xi)\,d\xi$$

$$= (2\pi)^{-n/2} \int_{\mathbb{R}^n} \sum_{\gamma\leq\beta} \binom{\beta}{\gamma} \xi^\gamma (D_\xi^\alpha e^{ix\cdot\xi})(D_x^{\beta-\gamma}\sigma)(x,\xi)\hat{\varphi}(\xi)\,d\xi$$

$$= (2\pi)^{-n/2}(-1)^{|\alpha|} \int_{\mathbb{R}^n} \sum_{\gamma\leq\beta} \binom{\beta}{\gamma} e^{ix\cdot\xi} D_\xi^\alpha \{(D_x^{\beta-\gamma}\sigma)(x,\xi)\xi^\gamma\hat{\varphi}(\xi)\}d\xi$$

$$= (2\pi)^{-n/2}(-1)^{|\alpha|} \int_{\mathbb{R}^n} \sum_{\gamma\leq\beta}\sum_{\delta\leq\alpha} \binom{\beta}{\gamma}\binom{\alpha}{\delta} e^{ix\cdot\xi} (D_\xi^{\alpha-\delta} D_x^{\beta-\gamma}\sigma)(x,\xi)$$

$$D_\xi^\delta(\xi^\gamma\hat{\varphi}(\xi))\,d\xi. \tag{6.11}$$

Using (6.11) and the fact that σ is a symbol, say $\sigma \in S^m$, we can find positive constants $C_{\alpha,\beta,\gamma,\delta}$, depending on α, β, γ and δ only, such that

$$\sup_{x\in\mathbb{R}^n} |x^\alpha (D^\beta (T_\sigma\varphi))(x)|$$

$$\leq (2\pi)^{-n/2} \sum_{\gamma\leq\beta}\sum_{\delta\leq\alpha} \binom{\beta}{\gamma}\binom{\alpha}{\delta} C_{\alpha,\beta,\gamma,\delta} \int_{\mathbb{R}^n} (1+|\xi|)^{m-|\alpha|+|\delta|}$$

$$|D_\xi^\delta(\xi^\gamma\hat{\varphi}(\xi))|\,d\xi. \tag{6.12}$$

Since $\varphi \in \mathcal{S}$, it follows from (6.12) that

$$\sup_{x \in \mathbb{R}^n} |x^\alpha (D^\beta (T_\sigma \varphi))(x)| < \infty,$$

and hence the proof is complete if we can justify the interchange of the order of differentiation and integration in (6.11). But, using the same argument as in the derivation of (6.12), we see that the last integral in (6.11) is absolutely convergent. This completes the proof. $\qquad \square$

Remark 6.8. In general, a pseudo-differential operator does not map $C_0^\infty(\mathbb{R}^n)$ into $C_0^\infty(\mathbb{R}^n)$. For a way of showing this, see Exercise 6.4.

An important notion in the theory of pseudo-differential operators is the asymptotic expansion of a symbol.

Definition 6.9. Let $\sigma \in S^m$. Suppose we can find $\sigma_j \in S^{m_j}$, where

$$m = m_0 > m_1 > m_2 > \cdots > m_j \to -\infty, \quad j \to \infty,$$

such that

$$\sigma - \sum_{j=0}^{N-1} \sigma_j \in S^{m_N} \tag{6.13}$$

for every positive integer N. Then we call $\sum_{j=0}^\infty \sigma_j$ an *asymptotic expansion* of σ and we write

$$\sigma \sim \sum_{j=0}^\infty \sigma_j.$$

An important result in this connection is the following theorem.

Theorem 6.10. *Let $m_0 > m_1 > m_2 > \cdots > m_j \to -\infty$ as $j \to \infty$. Suppose $\sigma_j \in S^{m_j}$. Then there exists a symbol $\sigma \in S^{m_0}$ such that*

$$\sigma \sim \sum_{j=0}^\infty \sigma_j.$$

Moreover, if τ is another symbol with the same asymptotic expansion, then

$$\sigma - \tau \in \cap_{m \in \mathbb{R}} S^m.$$

Proof Let $\psi \in C^\infty(\mathbb{R}^n)$ be such that

$$0 \leq \psi(\xi) \leq 1, \quad \xi \in \mathbb{R}^n,$$

$$\psi(\xi) = 0, \quad |\xi| \leq 1,$$

and

$$\psi(\xi) = 1, \quad |\xi| \geq 2.$$

That such a function exists will be proved at the end of this chapter. Let $\{\varepsilon_j\}$ be a sequence of positive numbers such that

$$1 > \varepsilon_0 > \varepsilon_1 > \varepsilon_2 > \cdots > \varepsilon_j \to 0$$

as $j \to \infty$. Define the function σ on $\mathbb{R}^n \times \mathbb{R}^n$ by

$$\sigma(x, \xi) = \sum_{j=0}^{\infty} \psi(\varepsilon_j \xi) \sigma_j(x, \xi), \quad x, \xi \in \mathbb{R}^n. \tag{6.14}$$

Note that for each $(x_0, \xi_0) \in \mathbb{R}^n \times \mathbb{R}^n$, there exists a neighborhood U of (x_0, ξ_0) and a positive integer N such that

$$\psi(\varepsilon_j \xi) \sigma_j(x, \xi) = 0$$

for all $(x, \xi) \in U$ and $j > N$. Hence $\sigma \in C^{\infty}(\mathbb{R}^n \times \mathbb{R}^n)$. Furthermore, for any $\varepsilon \in (0, 1]$ and nonzero multi-index α,

$$\psi(\varepsilon \xi) = 0, \quad |\xi| \leq \frac{1}{\varepsilon},$$

$$\psi(\varepsilon \xi) = 1, \quad |\xi| \geq \frac{2}{\varepsilon},$$

$$\partial_\xi^\alpha \{\psi(\varepsilon \xi)\} = \varepsilon^{|\alpha|}(\partial^\alpha \psi)(\varepsilon \xi) = 0, \quad |\xi| \leq \frac{1}{\varepsilon} \text{ or } |\xi| \geq \frac{2}{\varepsilon},$$

and

$$|\partial_\xi^\alpha \{\psi(\varepsilon \xi)\}| \leq C_\alpha \varepsilon^{|\alpha|}, \quad \xi \in \mathbb{R}^n,$$

where

$$C_\alpha = \sup_{\xi \in \mathbb{R}^n} |(\partial^\alpha \psi)(\xi)|.$$

If $\frac{1}{\varepsilon} \leq |\xi| \leq \frac{2}{\varepsilon}$, then $\varepsilon \leq \frac{2}{|\xi|} \leq \frac{4}{1+|\xi|}$. Hence, for any nonzero multi-index α, we have

$$|\partial_\xi^\alpha \{\psi(\varepsilon \xi)\}| \leq C_\alpha 4^{|\alpha|}(1 + |\xi|)^{-|\alpha|} = C_\alpha'(1 + |\xi|)^{-|\alpha|}, \quad \xi \in \mathbb{R}^n, \tag{6.15}$$

where $C_\alpha' = C_\alpha 4^{|\alpha|}$. It is obvious that (6.15) is also true for the zero multi-index α. Now, using (6.15), Leibniz's formula and the fact that $\sigma_j \in S^{m_j}$, we can find positive constants $C_{\alpha\gamma}$ and $C_{j,\beta,\gamma}$ such that

$$|D_\xi^\alpha D_x^\beta \{\psi(\varepsilon\xi)\sigma_j(x,\xi)\}|$$

$$= |D_\xi^\alpha \{\psi(\varepsilon\xi)(D_x^\beta \sigma_j)(x,\xi)\}|$$

$$= \left| \sum_{\gamma\leq\alpha} \binom{\alpha}{\gamma} (D_\xi^{\alpha-\gamma}\{\psi(\varepsilon\xi)\})(D_\xi^\gamma D_x^\beta \sigma_j)(x,\xi) \right|$$

$$\leq \sum_{\gamma\leq\alpha} \binom{\alpha}{\gamma} C_{\alpha,\gamma}(1+|\xi|)^{-|\alpha|+|\gamma|} C_{j,\beta,\gamma}(1+|\xi|)^{m_j-|\gamma|}$$

$$= \sum_{\gamma\leq\alpha} \binom{\alpha}{\gamma} C_{\alpha,\gamma} C_{j,\beta,\gamma}(1+|\xi|)^{m_j-|\alpha|}$$

$$= C_{j,\alpha,\beta}(1+|\xi|)^{-1}(1+|\xi|)^{m_j+1-|\alpha|} \tag{6.16}$$

for all $x,\xi \in \mathbb{R}^n$, where $C_{j,\alpha,\beta} = \sum_{\gamma\leq\alpha} \binom{\alpha}{\gamma} C_{\alpha,\gamma} C_{j,\beta,\gamma}$. Now, we choose ε_j such that

$$C_{j,\alpha,\beta}\varepsilon_j \leq 2^{-j} \tag{6.17}$$

for all multi-indices α and β such that $|\alpha+\beta| \leq j$. By the definition of ψ, we have

$$\psi(\varepsilon_j\xi) = 0 \tag{6.18}$$

whenever $1 + |\xi| \leq \varepsilon_j^{-1}$. Hence, by (6.16), (6.17) and (6.18),

$$|D_\xi^\alpha D_x^\beta \{\psi(\varepsilon_j\xi)\sigma_j(x,\xi)\}| \leq 2^{-j}(1+|\xi|)^{m_j+1-|\alpha|} \tag{6.19}$$

whenever $x,\xi \in \mathbb{R}^n$ and $|\alpha+\beta| \leq j$. Now, for any two multi-indices α_0 and β_0, we take j_0 so large that $j_0 \geq |\alpha_0 + \beta_0|$ and $m_{j_0} + 1 \leq m_0$. Write

$$\sigma(x,\xi) = \sum_{j=0}^{j_0-1} \psi(\varepsilon_j\xi)\sigma_j(x,\xi) + \sum_{j=j_0}^{\infty} \psi(\varepsilon_j\xi)\sigma_j(x,\xi)$$

$$= I(x,\xi) + J(x,\xi). \tag{6.20}$$

Since $I(x,\xi)$ is a finite sum, it follows that $I \in S^{m_0}$. By (6.19),

$$|(D_\xi^{\alpha_0} D_x^{\beta_0} J)(x,\xi)| \leq \sum_{j=j_0}^{\infty} |D_\xi^{\alpha_0} D_x^{\beta_0}\{\psi(\varepsilon_j\xi)\sigma_j(x,\xi)\}|$$

$$\leq \sum_{j=j_0}^{\infty} 2^{-j}(1+|\xi|)^{m_j+1-|\alpha_0|}$$

$$\leq \sum_{j=j_0}^{\infty} 2^{-j}(1+|\xi|)^{m_0-|\alpha_0|}$$

$$= 2^{-j_0+1}(1+|\xi|)^{m_0-|\alpha_0|}. \tag{6.21}$$

So, J is also in S^{m_0}. Hence, by (6.20), $\sigma \in S^{m_0}$. We need to verify condition (6.13). To do this, we write

$$\sigma(x, \xi) - \sum_{j=0}^{N-1} \sigma_j(x, \xi)$$

$$= \sum_{j=0}^{\infty} \psi(\varepsilon_j \xi) \sigma_j(x, \xi) - \sum_{j=0}^{N-1} \sigma_j(x, \xi)$$

$$= \sum_{j=0}^{N-1} \{\psi(\varepsilon_j \xi) - 1\} \sigma_j(x, \xi) + \sum_{j=N}^{\infty} \psi(\varepsilon_j \xi) \sigma_j(x, \xi).$$

As before, we can show that

$$\sum_{j=N}^{\infty} \psi(\varepsilon_j \xi) \sigma_j(x, \xi) \in S^{m_N}.$$

Since

$$\psi(\varepsilon_j \xi) - 1 = 0, \quad j \le N - 1,$$

if $|\xi| \ge \frac{2}{\varepsilon_{N-1}}$, it follows from Exercise 6.3 that

$$\sum_{j=0}^{N-1} \{\psi(\varepsilon_j \xi) - 1\} \sigma_j(x, \xi) \in \cap_{m \in \mathbb{R}} S^m$$

and consequently

$$\sigma - \sum_{j=0}^{N-1} \sigma_j \in S^{m_N}.$$

Finally, if τ is another symbol such that $\tau \sim \sum_{j=0}^{\infty} \sigma_j$, then

$$\sigma - \tau = \left[\sigma - \sum_{j=0}^{N-1} \sigma_j \right] - \left[\tau - \sum_{j=0}^{N-1} \sigma_j \right] \in S^{m_N}$$

for every positive integer N. Since $m_N \to -\infty$ as $N \to \infty$, it follows that

$$\sigma - \tau \in \cap_{m \in \mathbb{R}} S^m.$$

This completes the proof of the theorem. $\qquad\square$

We have used the following result in the proof of Theorem 6.10.

Proposition 6.11. *There exists a function* $\psi \in C^{\infty}(\mathbb{R}^n)$ *such that*

$$0 \le \psi(\xi) \le 1, \quad \xi \in \mathbb{R}^n,$$

$$\psi(\xi) = 0, \quad |\xi| \le 1,$$

and

$$\psi(\xi) = 1, \quad |\xi| \ge 2.$$

Proof We need only construct a function $\varphi_0 \in C_0^\infty(\mathbb{R}^n)$ such that

$$0 \leq \varphi_0(\xi) \leq 1, \quad \xi \in \mathbb{R}^n,$$

$$\varphi_0(\xi) = 1, \quad |\xi| \leq 1,$$

and

$$\varphi_0(\xi) = 0, \quad |\xi| \geq 2.$$

For then the function $\psi = 1 - \varphi_0$ will satisfy all the conditions of Proposition 6.11. To construct φ_0, let f be any continuous function on $\mathbb{R}^n$ such that

$$0 \leq f(t) \leq 1, \quad t \in \mathbb{R}^n,$$

$$f(t) = 1, \quad |t| \leq \frac{3}{2},$$

and

$$f(t) = 0, \quad |t| \geq \frac{7}{4}.$$

Let $\varphi \in C_0^\infty(\mathbb{R}^n)$ be a real-valued and nonnegative function such that

$$\varphi(s) = 0, \quad |s| \geq \frac{1}{4}.$$

and

$$\int_{|s| \leq \frac{1}{4}} \varphi(s)\, ds = 1. \tag{6.22}$$

That such a function φ exists is an immediate consequence of Exercise 3.2. Let $\varphi_0 = f * \varphi$. Then, by Propositions 3.6 and 3.7, $\varphi_0 \in C_0^\infty(\mathbb{R}^n)$. Furthermore, using the location of the supports of f and φ, we see that

$$\varphi_0(t) = 0, \quad |t| \geq 2.$$

Finally, for $|t| \leq 1$,

$$\varphi_0(t) = \int_{\mathbb{R}^n} f(t - s)\varphi(s)\, ds = \int_{|s| \leq \frac{1}{4}} f(t - s)\varphi(s)\, ds. \tag{6.23}$$

Since, for $|t| \leq 1$ and $|s| \leq \frac{1}{4}$, we have $|s - t| \leq \frac{5}{4}$ and hence $f(t - s) = 1$. Therefore, by (6.22) and (6.23),

$$\varphi_0(t) = 1, \quad |t| \leq 1.$$

Again, by (6.22) and (6.23), we can prove easily that

$$0 \leq \varphi_0(t) \leq 1, \quad t \in \mathbb{R}^n.$$

This completes the proof of the proposition. $\qquad\qquad\qquad\square$

Remark 6.12. For another proof of Proposition 6.11, see Exercise 6.6.

Exercises

6.1. Prove that if $\sigma \in S^{m_1}$ and $\tau \in S^{m_2}$, then $\sigma\tau \in S^{m_1+m_2}$.

6.2. Prove that if $\sigma \in S^m$, then $D_x^\alpha D_\xi^\beta \sigma \in S^{m-|\beta|}$ for all multi-indices α and β.

6.3. Let σ be any symbol and φ any function in S. Prove that the function τ defined by

$$\tau(x,\xi) = \sigma(x,\xi)\varphi(\xi), \quad x,\xi \in \mathbb{R}^n,$$

is a symbol in $\cap_{m\in\mathbb{R}}S^m$.

6.4. Let σ be the symbol defined by

$$\sigma(\xi) = e^{-|\xi|^2/2}$$

for all $\xi \in \mathbb{R}^n$. Show that the pseudo-differential operator T_σ does not map $C_0^\infty(\mathbb{R}^n)$ into $C_0^\infty(\mathbb{R}^n)$.

6.5. Let α and δ be any two multi-indices. Prove that for all $\xi \in \mathbb{R}^n$,

$$\partial_\xi^\delta \xi^\alpha = \begin{cases} \delta!\binom{\alpha}{\delta}\xi^{\alpha-\delta}, & \delta \leq \alpha, \\ 0, & \text{otherwise.} \end{cases}$$

6.6. (i) Let $\varphi \in C_0^\infty(\mathbb{R})$ be such that $\varphi(x) \geq 0$ for $x \in \mathbb{R}$, $\varphi(x) = 0$ for $x \notin [-2,-1]$, and

$$\int_{-\infty}^\infty \varphi(x)\,dx = 1.$$

Define a function χ on $(-\infty,0]$ by

$$\chi(\xi) = \int_{-\infty}^\xi \varphi(x)\,dx, \quad \xi \in (-\infty,0].$$

Prove that χ is infinitely differentiable on $(-\infty,0)$,

$$0 \leq \chi(\xi) \leq 1, \quad \xi \in (-\infty,0],$$

$$\chi(\xi) = 0, \quad \xi \leq -2,$$

and

$$\chi(\xi) = 1, \quad \xi \in [-1,0].$$

(ii) Let $\varphi \in C_0^\infty(\mathbb{R})$ be such that $\varphi(x) \geq 0$ for $x \in \mathbb{R}$, $\varphi(x) = 0$ for $x \notin [1,2]$ and

$$\int_{-\infty}^\infty \varphi(x)\,dx = 1.$$

Define a function χ on $[0, \infty)$ by

$$\chi(\xi) = \int_{\xi}^{\infty} \varphi(x)\, dx, \quad \xi \in [0, \infty).$$

Prove that χ is infinitely differentiable on $(0, \infty)$,

$$0 \le \chi(\xi) \le 1, \quad \xi \in [0, \infty),$$

$$\chi(\xi) = 0, \quad \xi \ge 2,$$

and

$$\chi(\xi) = 1, \quad \xi \in [0, 1].$$

(iii) Construct a function $\psi \in C^{\infty}(\mathbb{R}^n)$ such that

$$0 \le \psi(\xi) \le 1, \quad \xi \in \mathbb{R}^n,$$

$$\psi(\xi) = 0, \quad |\xi| \le 1,$$

and

$$\psi(\xi) = 1, \quad |\xi| \ge 2.$$

6.7. Let $\sigma \in C^{\infty}(\mathbb{R}^n \times \mathbb{R}^n)$ and let $m \in (-\infty, \infty)$. Prove that $\sigma \in S^m$ if and only if for all multi-indices α and β, there exist positive constants $C_{\alpha, \beta}$ and $R_{\alpha, \beta}$ such that

$$|(D_x^{\alpha} D_{\xi}^{\beta} \sigma)(x, \xi)| \le C_{\alpha, \beta}(1 + |\xi|)^{m - |\beta|}, \quad |\xi| \ge R_{\alpha, \beta}.$$

Chapter 7

A Partition of Unity and Taylor's Formula

It is convenient to devote a chapter to several technical results which will be of particular importance for us in the next two chapters. In Theorem 7.1 we construct a partition of unity. Then we use this partition of unity to decompose a symbol $\sigma(x, \xi)$ into a family $\{\sigma_k(x, \xi)\}$ of symbols with compact support in the ξ variable. We are able to obtain good estimates on the partial Fourier transforms (with respect to the ξ variable) of all the symbols $\sigma_k(x, \xi)$. The precise estimates are given in Theorem 7.2. In Theorem 7.3 we prove a multi-dimensional version of Taylor's formula with integral remainder.

We begin with the construction of a partition of unity.

Theorem 7.1. *There is a sequence $\{\varphi_k\}_{k=0}^{\infty}$ of functions in $C_0^{\infty}(\mathbb{R}^n)$ such that*

(i) $0 \leq \varphi_k(\xi) \leq 1, \quad \xi \in \mathbb{R}^n, \, k = 0, 1, 2, \ldots,$

(ii) $\sum_{k=0}^{\infty} \varphi_k(\xi) = 1, \quad \xi \in \mathbb{R}^n,$

(iii) *at each $\xi \in \mathbb{R}^n$, at least one and at most three of the $\varphi_k's$ are nonzero,*

(iv) $\operatorname{supp}(\varphi_0) \subseteq \{\xi \in \mathbb{R}^n : |\xi| \leq 2\},$

(v) $\operatorname{supp}(\varphi_k) \subseteq \{\xi \in \mathbb{R}^n : 2^{k-2} \leq |\xi| \leq 2^{k+1}\}, \quad k = 1, 2, \ldots,$

(vi) *for each multi-index α, there is a constant $A_\alpha > 0$ such that*

$$\sup_{\xi \in \mathbb{R}^n} |(\partial^\alpha \varphi_k)(\xi)| \leq A_\alpha 2^{-k|\alpha|}, \quad k = 0, 1, 2, \ldots.$$

Proof We pick ψ_0 to be any function in $C_0^{\infty}(\mathbb{R}^n)$ such that

$$0 \leq \psi_0(\xi) \leq 1, \quad \xi \in \mathbb{R}^n,$$

$$\psi_0(\xi) = 1, \quad |\xi| \leq 1,$$

and

$$\psi_0(\xi) = 0, \quad |\xi| > 2.$$

For the existence of such a function, see Proposition 6.11 and Remark 6.12. Let ψ be any other function in $C_0^\infty(\mathbb{R}^n)$ such that

$$0 \leq \psi(\xi) \leq 1, \quad \xi \in \mathbb{R}^n,$$

$$\psi(\xi) = 1, \quad 1 \leq |\xi| \leq 2,$$

and

$$\psi(\xi) = 0, \quad |\xi| < \frac{1}{2} \text{ or } |\xi| > 4.$$

The proof that such a function ψ exists is left as Exercise 7.1. Now, for $k = 1, 2, \ldots$, define ψ_k by

$$\psi_k(\xi) = \psi\left(\frac{\xi}{2^{k-1}}\right), \quad \xi \in \mathbb{R}^n,$$

and then define Ψ by

$$\Psi(\xi) = \sum_{k=0}^\infty \psi_k(\xi), \quad \xi \in \mathbb{R}^n.$$

Obviously, we have

$$\operatorname{supp}(\psi_0) \subseteq \{\xi \in \mathbb{R}^n : |\xi| \leq 2\}$$

and

$$\operatorname{supp}(\psi_k) \subseteq \{\xi \in \mathbb{R}^n : 2^{k-2} \leq |\xi| \leq 2^{k+1}\}$$

for $k = 1, 2, \ldots$. For each $\xi \in \mathbb{R}^n$, the series defining $\Psi(\xi)$ contains at most three nonzero consecutive terms. This implies that for each $\xi_0 \in \mathbb{R}^n$, there exists a neighborhood U of ξ_0 and a positive integer N such that $\psi_k(\xi) = 0$ for all $\xi \in U$ and $k > N$. Hence $\Psi \in C^\infty(\mathbb{R}^n)$. It is easy to see that for each $\xi \in \mathbb{R}^n$, the series defining $\Psi(\xi)$ contains at least one nonzero term. Hence $\Psi(\xi) \neq 0$ for all $\xi \in \mathbb{R}^n$. Now, for $k = 0, 1, 2, \ldots$, we define φ_k by

$$\varphi_k(\xi) = \frac{\psi_k(\xi)}{\Psi(\xi)}, \quad \xi \in \mathbb{R}^n.$$

It is easy to see that $\varphi_k \in C_0^\infty(\mathbb{R}^n)$ for $k = 0, 1, 2, \ldots$ and the sequence $\{\varphi_k\}_{k=0}^\infty$ satisfies the first five conditions. Hence it remains to prove that the last condition is true for φ_k, $k = 1, 2, \ldots$. To do this, we note that for each multi-index α, we have, by Leibniz's formula,

$$(\partial^\alpha \varphi_k)(\xi)$$

$$= \sum_{\beta \leq \alpha} \binom{\alpha}{\beta} \left\{\partial^\beta\left(\frac{1}{\Psi}\right)(\xi)\right\} (\partial^{\alpha-\beta}\psi_k)(\xi)$$

$$= \sum_{\beta \leq \alpha} \binom{\alpha}{\beta} \left\{\partial^\beta\left(\frac{1}{\Psi}\right)(\xi)\right\} (\partial^{\alpha-\beta}\psi)\left(\frac{\xi}{2^{k-1}}\right) 2^{-(k-1)|\alpha-\beta|}. \quad (7.1)$$

Now, for each multi-index β, by formula (1.4),

$$\partial^\beta \left(\frac{1}{\Psi}\right) = \sum_{\beta^{(1)}+\cdots+\beta^{(l)}=\beta} C_{\beta^{(1)},\ldots,\beta^{(l)}} \frac{(\partial^{\beta^{(1)}}\Psi)\cdots(\partial^{\beta^{(l)}}\Psi)}{\Psi^{l+1}}, \tag{7.2}$$

where $C_{\beta^{(1)},\ldots,\beta^{(l)}}$ is a constant and the sum is taken over all possible multi-indices $\beta^{(1)},\ldots,\beta^{(l)}$ which partition β. Next, for every multi-index γ, there exists a constant $C_\gamma > 0$ such that

$$|(\partial^\gamma\Psi)(\xi)| \le C_\gamma 2^{-k|\gamma|} \tag{7.3}$$

for all $\xi \in \text{supp}(\varphi_k)$ and $k = 0,1,2,\ldots$. Let us assume (7.3) for a moment. Then, by (7.2) and (7.3), there exist positive constants $C_{\beta^{(1)}},\ldots,C_{\beta^{(l)}}$ such that

$$\left|\partial^\beta\left(\frac{1}{\Psi}\right)(\xi)\right| \le \sum_{\beta^{(1)}+\cdots+\beta^{(l)}=\beta} |C_{\beta^{(1)},\ldots,\beta^{(l)}}| \frac{C_{\beta^{(1)}}\cdots C_{\beta^{(l)}}2^{-k|\beta|}}{|\Psi(\xi)|^{l+1}}$$

$$\le C_\beta' 2^{-k|\beta|} \tag{7.4}$$

for all $\xi \in \text{supp}(\varphi_k)$, where

$$C_\beta' = \sum_{\beta^{(1)}+\cdots+\beta^{(l)}=\beta} |C_{\beta^{(1)},\ldots,\beta^{(l)}}||C_{\beta^{(1)}}\cdots C_{\beta^{(l)}}|.$$

Hence, by (7.1) and (7.4), there exist positive constants $C_{\alpha,\beta}$ such that

$$|(\partial^\alpha\varphi_k)(\xi)| \le \sum_{\beta\le\alpha} \binom{\alpha}{\beta} C_\beta' 2^{-k|\beta|} C_{\alpha,\beta} 2^{-(k-1)|\alpha-\beta|} = A_\alpha 2^{-k|\alpha|}$$

for all $k = 0,1,2,\ldots$ and all $\xi \in \mathbb{R}^n$, where

$$A_\alpha = \sum_{\beta\le\alpha} \binom{\alpha}{\beta} C_\beta' C_{\alpha,\beta} 2^{|\alpha-\beta|}.$$

It remains to prove (7.3). To do this, we consider three cases.

Case 1: Suppose $k = 0$. Then for all $\xi \in \text{supp}(\varphi_0)$, ξ is in $\text{supp}(\psi_j)$ for some $j = 0,1,2$ and hence

$$\Psi(\xi) = \psi_0(\xi) + \sum_{j=1}^{2} \psi_j(\xi).$$

Therefore

$$(\partial^\gamma\Psi)(\xi) = (\partial^\gamma\psi_0)(\xi) + \sum_{j=1}^{2}(\partial^\gamma\psi)\left(\frac{\xi}{2^{j-1}}\right) 2^{-(j-1)|\gamma|}$$

for all $\xi \in \text{supp}(\varphi_0)$. Thus, there is a constant $C_\gamma > 0$ such that

$$|(\partial^\gamma \Psi)(\xi)| \le C_\gamma, \quad \xi \in \text{supp}(\varphi_0). \tag{7.5}$$

Case 2: Suppose $k = 1$. Then for all $\xi \in \text{supp}(\varphi_1)$, ξ is in $\text{supp}(\psi_j)$ for some $j = 0, 1, 2, 3$ and hence

$$\Psi(\xi) = \psi_0(\xi) + \sum_{j=1}^{3} \psi_j(\xi).$$

Therefore

$$(\partial^\gamma \Psi)(\xi) = (\partial^\gamma \psi_0)(\xi) + \sum_{j=1}^{3} (\partial^\gamma \psi) \left(\frac{\xi}{2^{j-1}} \right) 2^{-(j-1)|\gamma|}$$

for all $\xi \in \text{supp}(\varphi_1)$. Thus, there is a constant $C'_\gamma > 0$ such that

$$\begin{aligned}
|(\partial^\gamma \Psi)(\xi)| &\le C'_\gamma [2^{|\gamma|} + 1 + 2^{-|\gamma|} + 2^{-2|\gamma|}] \\
&= C'_\gamma [2^{2|\gamma|} + 2^{|\gamma|} + 1 + 2^{-|\gamma|}] 2^{-|\gamma|}
\end{aligned} \tag{7.6}$$

for all $\xi \in \text{supp}(\varphi_1)$.

Case 3: Suppose $k \ge 2$. Then for all $\xi \in \text{supp}(\varphi_k)$, ξ is in $\text{supp}(\psi_j)$ for some $j = k - 2, k - 1, k, k + 1, k + 2$ and hence

$$\Psi(\xi) = \sum_{j=k-2}^{k+2} \psi_j(\xi).$$

Therefore

$$(\partial^\gamma \Psi)(\xi) = \sum_{j=k-2}^{k+2} (\partial^\gamma \psi) \left(\frac{\xi}{2^{j-1}} \right) 2^{-(j-1)|\gamma|}$$

for all $\xi \in \text{supp}(\varphi_k)$. Thus, there is a constant $C''_\gamma > 0$ such that

$$\begin{aligned}
&|(\partial^\gamma \Psi)(\xi)| \\
&\le C''_\gamma [2^{-(k-3)|\gamma|} + 2^{-(k-2)|\gamma|} + 2^{-(k-1)|\gamma|} + 2^{-k|\gamma|} + 2^{-(k+1)|\gamma|}] \\
&= C''_\gamma [2^{3|\gamma|} + 2^{2|\gamma|} + 2^{|\gamma|} + 1 + 2^{-|\gamma|}] 2^{-k|\gamma|}
\end{aligned} \tag{7.7}$$

for all $\xi \in \text{supp}(\varphi_k)$. Hence, by (7.5), (7.6) and (7.7), (6.3) follows and the proof of Theorem 7.1 is complete. $\square$

Let $\sigma \in S^m$. For $k = 0, 1, 2, \ldots$, we write

$$\sigma_k(x, \xi) = \sigma(x, \xi) \varphi_k(\xi) \tag{7.8}$$

for all $x, \xi \in \mathbb{R}^n$ and

$$K_k(x, z) = (2\pi)^{-n/2} \int_{\mathbb{R}^n} e^{iz \cdot \xi} \sigma_k(x, \xi)\, d\xi \qquad (7.9)$$

for all $x, z \in \mathbb{R}^n$, where $\{\varphi_k\}$ is the partition of unity constructed in Theorem 7.1.

Theorem 7.2. *For all nonnegative integers N, and multi-indices α and β, there exists a constant A, depending on m, n, N, α and β only, such that*

$$\int_{\mathbb{R}^n} |z|^N |(\partial_x^\beta \partial_z^\alpha K_k)(x, z)|\, dz \leq A 2^{(m+|\alpha|-N)k}$$

for all $k = 0, 1, 2, \dots$.

In the proof of Theorem 7.2, we make use of the following inequality given by

$$|z|^{2N} \leq n^N \sum_{|\gamma|=N} |z^\gamma|^2, \quad z \in \mathbb{R}^n. \qquad (7.10)$$

The proof of (7.10) is left as an exercise. See Exercise 7.2.

Proof of Theorem 7.2 Let γ be any multi-index. Then, by (7.8), (7.9), Plancherel's theorem, Proposition 4.2, Leibniz's formula, parts (iv) and (v) of Theorem 7.1,

$$\int_{\mathbb{R}^n} |z^\gamma (\partial_x^\beta \partial_z^\alpha K_k)(x, z)|^2 dz$$

$$= \int_{\mathbb{R}^n} |\partial_\xi^\gamma \{\xi^\alpha (\partial_x^\beta \sigma_k)(x, \xi)\}|^2 d\xi$$

$$= \int_{W_k} \left| \sum_{\gamma' \leq \gamma} \binom{\gamma}{\gamma'} \partial_\xi^{\gamma'} \{\xi^\alpha (\partial_x^\beta \sigma)(x, \xi)\} (\partial_\xi^{\gamma - \gamma'} \varphi_k)(\xi) \right|^2 d\xi, \qquad (7.11)$$

where

$$W_0 = \{\xi \in \mathbb{R}^n : |\xi| \leq 2\} \qquad (7.12)$$

and

$$W_k = \{\xi \in \mathbb{R}^n : 2^{k-2} \leq |\xi| \leq 2^{k+1}\} \qquad (7.13)$$

for $k = 1, 2, \dots$. Hence, by (7.11), part (vi) of Theorem 7.1 and the fact that $\xi^\alpha (\partial_x^\beta \sigma)$ is a symbol in $S^{m+|\alpha|}$, we get positive constants $C_{\alpha, \beta, \gamma'}$ and

$C_{\gamma,\gamma'}$ such that

$$\int_{\mathbb{R}^n} |z^\gamma (\partial_x^\beta \partial_z^\alpha K_k)(x,z)|^2 dz$$

$$\leq \int_{W_k} \left\{ \sum_{\gamma' \leq \gamma} \binom{\gamma}{\gamma'} C_{\alpha,\beta,\gamma'} (1+|\xi|)^{m+|\alpha|-|\gamma'|} C_{\gamma,\gamma'} 2^{-k|\gamma-\gamma'|} \right\}^2 d\xi.$$

(7.14)

Hence, by (7.12), (7.13) and (7.14),

$$\int_{\mathbb{R}^n} |z^\gamma (\partial_x^\beta \partial_z^\alpha K_k)(x,z)|^2 dz$$

$$\leq \int_{W_k} \left\{ \sum_{\gamma' \leq \gamma} \binom{\gamma}{\gamma'} C_{\alpha,\beta,\gamma'} 2^{(k+2)(m+|\alpha|-|\gamma'|)} C_{\gamma,\gamma'} 2^{-k|\gamma-\gamma'|} \right\}^2 d\xi$$

$$= 2^{k(2m+2|\alpha|-2|\gamma|)} \int_{W_k} \left\{ \sum_{\gamma' \leq \gamma} \binom{\gamma}{\gamma'} C_{\alpha,\beta,\gamma'} C_{\gamma,\gamma'} 2^{2(m+|\alpha|-|\gamma'|)} \right\}^2 d\xi$$

$$\leq 2^{k(n+2m+2|\alpha|-2|\gamma|)} C_n 2^n \left\{ \sum_{\gamma' \leq \gamma} \binom{\gamma}{\gamma'} C_{\alpha\,\beta\,\gamma'} C_{\gamma,\gamma'} 2^{2(m+|\alpha|-|\gamma'|)} \right\}^2,$$

(7.15)

where C_n is the constant with the property that the volume of the ball in $\mathbb{R}^n$ with radius r is equal to $C_n r^n$. Let $A_{\alpha,\beta,\gamma,m,n}$ be defined by

$$A_{\alpha,\beta,\gamma,m,n} = C_n 2^n \left\{ \sum_{\gamma' \leq \gamma} \binom{\gamma}{\gamma'} C_{\alpha,\beta,\gamma'} C_{\gamma,\gamma'} 2^{2(m+|\alpha|-|\gamma'|)} \right\}^2.$$

Then (7.15) becomes

$$\int_{\mathbb{R}^n} |z^\gamma (\partial_x^\beta \partial_z^\alpha K_k)(x,z)|^2 dz \leq A_{\alpha,\beta,\gamma,m,n} 2^{k(n+2m+2|\alpha|-2|\gamma|)}.$$ (7.16)

Let N be any nonnegative integer. Then, by (7.10) and (7.16),

$$\int_{\mathbb{R}^n} |z|^{2N} |(\partial_x^\beta \partial_z^\alpha K_k)(x,z)|^2 dz$$

$$\leq n^N \sum_{|\gamma|=N} \int_{\mathbb{R}^n} |z^\gamma (\partial_x^\beta \partial_z^\alpha K_k)(x,z)|^2 dz$$

$$\leq A_1 2^{k(n+2m+2|\alpha|-2N)}$$

for all $k = 0, 1, 2, \ldots$, where

$$A_1 = n^N \sum_{|\gamma|=N} A_{\alpha,\beta,\gamma,m,n}.$$

By taking square roots, we get

$$\left\{ \int_{\mathbb{R}^n} |z|^{2N} |(\partial_x^\beta \partial_z^\alpha K_k)(x,z)|^2 dz \right\}^{1/2} \leq A_2 2^{((n/2)+m+|\alpha|-N)k} \qquad (7.17)$$

for $k = 0, 1, 2, \ldots$, where $A_2 = A_1^{1/2}$. Now write

$$\int_{\mathbb{R}^n} |z|^N |(\partial_x^\beta \partial_z^\alpha K_k)(x,z)| \, dz = \int_{|z| \leq 2^{-k}} + \int_{|z| > 2^{-k}}. \qquad (7.18)$$

By (7.17) and the Cauchy–Schwarz inequality, there is a constant $A_3 > 0$, depending on m, n, N, α and β only, such that

$$\int_{|z| \leq 2^{-k}} \leq \left\{ \int_{\mathbb{R}^n} |z|^{2N} |(\partial_x^\beta \partial_z^\alpha K_k)(x,z)|^2 dz \right\}^{1/2} \left\{ \int_{|z| \leq 2^{-k}} dz \right\}^{1/2}$$

$$\leq A_3 2^{((n/2)+m+|\alpha|-N)k} 2^{-nk/2}$$

$$= A_3 2^{(m+|\alpha|-N)k} \qquad (7.19)$$

for $k = 0, 1, 2, \ldots$. Next, by (7.17) and the Cauchy–Schwarz inequality again, there is another constant $A_4 > 0$, depending on m, n, N, α and β only, such that

$$\int_{|z| > 2^{-k}}$$

$$\leq \left\{ \int_{\mathbb{R}^n} |z|^{2(N+n)} |(\partial_x^\beta \partial_z^\alpha K_k)(x,z)|^2 dz \right\}^{1/2} \left\{ \int_{|z| > 2^{-k}} |z|^{-2n} dz \right\}^{1/2}$$

$$\leq A_4 2^{((n/2)+m+|\alpha|-N-n)k} \left\{ \int_{S^{n-1}} \int_{2^{-k}}^\infty r^{-2n} r^{n-1} dr \, d\sigma \right\}^{1/2},$$

where $d\sigma$ is the surface measure on the unit sphere S^{n-1}. Hence

$$\int_{|z| > 2^{-k}} \leq A_4 2^{((n/2)+m+|\alpha|-N-n)k} |S^{n-1}|^{1/2} n^{-1/2} 2^{nk/2}$$

$$= A_5 2^{(m+|\alpha|-N)k} \qquad (7.20)$$

for $k = 0, 1, 2, \ldots$, where $A_5 = A_4 |S^{n-1}|^{1/2} n^{-1/2}$ and $|S^{n-1}|$ is the surface area of S^{n-1}. Hence, by (7.18), (7.19) and (7.20), we complete the proof of Theorem 7.2. $\qquad \square$

The following version of Taylor's formula with integral remainder plays an important role in the study of pseudo-differential operators.

Theorem 7.3. *Let $f \in C^\infty(\mathbb{R}^n)$. Then for all positive integers N,*

$$
\begin{aligned}
&f(\xi + \eta) \\
&= \sum_{|\alpha| < N} \frac{(\partial^\alpha f)(\xi)}{\alpha!} \eta^\alpha + N \sum_{|\gamma| = N} \frac{\eta^\gamma}{\gamma!} \int_0^1 (1 - \theta)^{N-1} (\partial^\gamma f)(\xi + \theta\eta)\, d\theta
\end{aligned}
\tag{7.21}
$$

for all $\xi, \eta \in \mathbb{R}^n$.

Proof The proof is by induction on N. For $N = 1$, we need to prove that

$$
f(\xi + \eta) = f(\xi) + \sum_{|\gamma|=1} \frac{\eta^\gamma}{\gamma!} \int_0^1 (\partial^\gamma f)(\xi + \theta\eta)\, d\theta
\tag{7.22}
$$

for all $\xi, \eta \in \mathbb{R}^n$. To do this, we define a function $\varphi : \mathbb{R} \to \mathbb{C}$ by

$$
\varphi(x) = f(\xi + x\eta), \quad x \in \mathbb{R}.
\tag{7.23}
$$

Then, by the fundamental theorem of calculus,

$$
\varphi(1) = \varphi(0) + \int_0^1 \varphi'(t)\, dt.
\tag{7.24}
$$

Hence, by (7.23), (7.24) and the chain rule, we get the formula

$$
f(\xi + \eta) = f(\xi) + \sum_{j=1}^n \int_0^1 (\partial_j f)(\xi + t\eta)\eta_j\, dt, \quad \xi, \eta \in \mathbb{R}^n,
$$

which is exactly the same as (7.22). We now assume that (7.21) is true for the positive integer N. Then, by the induction hypothesis, we have

$$
\begin{aligned}
&f(\xi + \eta) - \sum_{|\alpha| < N+1} \frac{(\partial^\alpha f)(\xi)}{\alpha!} \eta^\alpha \\
&= f(\xi + \eta) - \sum_{|\alpha| < N} \frac{(\partial^\alpha f)(\xi)}{\alpha!} \eta^\alpha - \sum_{|\alpha| = N} \frac{(\partial^\alpha f)(\xi)}{\alpha!} \eta^\alpha \\
&= N \sum_{|\gamma| = N} \frac{\eta^\gamma}{\gamma!} \int_0^1 (1 - \theta)^{N-1} (\partial^\gamma f)(\xi + \theta\eta)\, d\theta - \sum_{|\alpha| = N} \frac{(\partial^\alpha f)(\xi)}{\alpha!} \eta^\alpha
\end{aligned}
\tag{7.25}
$$

for all $\xi, \eta \in \mathbb{R}^n$. Now, using the fact that

$$
N \int_0^1 (1 - \theta)^{N-1}\, d\theta = 1
$$

and the formula (7.21) for $N = 1$, the formula (7.25) becomes

$$f(\xi + \eta) - \sum_{|\alpha| < N+1} \frac{(\partial^\alpha f)(\xi)}{\alpha!} \eta^\alpha$$

$$= \sum_{|\gamma|=N} \frac{\eta^\gamma}{\gamma!} N \int_0^1 (1 - \theta)^{N-1} \{(\partial^\gamma f)(\xi + \theta\eta) - (\partial^\gamma f)(\xi)\} d\theta$$

$$= \sum_{|\gamma|=N} \frac{\eta^\gamma}{\gamma!} N \int_0^1 (1 - \theta)^{N-1} \sum_{|\delta|=1} \frac{(\theta\eta)^\delta}{\delta!} \left\{ \int_0^1 (\partial^{\gamma+\delta} f)(\xi + \varphi\theta\eta) \, d\varphi \right\} d\theta.$$

$$(7.26)$$

Let $\rho = \varphi\theta$ in (7.26). Then

$$f(\xi + \eta) - \sum_{|\alpha| < N+1} \frac{(\partial^\alpha f)(\xi)}{\alpha!} \eta^\alpha$$

$$= \sum_{|\gamma|=N} \sum_{|\delta|=1} \frac{\eta^{\gamma+\delta}}{\gamma!\delta!} N \int_0^1 (1 - \theta)^{N-1} \left\{ \int_0^\theta (\partial^{\gamma+\delta} f)(\xi + \rho\eta) \, d\rho \right\} d\theta.$$

$$(7.27)$$

By interchanging the order of integration in (7.27), we have

$$f(\xi + \eta) - \sum_{|\alpha| < N+1} \frac{(\partial^\alpha f)(\xi)}{\alpha!} \eta^\alpha$$

$$= \sum_{|\gamma|=N} \sum_{|\delta|=1} \frac{\eta^{\gamma+\delta}}{\gamma!\delta!} N \int_0^1 (\partial^{\gamma+\delta} f)(\xi + \rho\eta) \left\{ \int_\rho^1 (1 - \theta)^{N-1} d\theta \right\} d\rho$$

$$= \sum_{|\gamma|=N} \sum_{|\delta|=1} \frac{\eta^{\gamma+\delta}}{\gamma!\delta!} \int_0^1 (\partial^{\gamma+\delta} f)(\xi + \rho\eta)(1 - \rho)^N d\rho. \qquad (7.28)$$

For all multi-indices γ and δ with $|\gamma| = N$ and $|\delta| = 1$, we have

$$\frac{(\gamma + \delta)!}{\gamma!\delta!} = \gamma \cdot \delta + 1, \qquad (7.29)$$

where $\gamma \cdot \delta$ is the inner product of γ and δ. Hence, by (7.28) and (7.29),

$$f(\xi + \eta) - \sum_{|\alpha| < N+1} \frac{(\partial^\alpha f)(\xi)}{\alpha!} \eta^\alpha$$

$$= \sum_{|\gamma|=N} \sum_{|\delta|=1} \frac{\eta^{\gamma+\delta}}{(\gamma + \delta)!} (\gamma \cdot \delta + 1) \int_0^1 (\partial^{\gamma+\delta} f)(\xi + \rho\eta)(1 - \rho)^N d\rho.$$

$$(7.30)$$

For each multi-index α of length $N + 1$, we can write α in the form $\gamma + \delta$, where $|\gamma| = N$ and $|\delta| = 1$ by picking

$$\gamma = (\alpha_1, \alpha_2, \ldots, \alpha_j - 1, \ldots, \alpha_n) \tag{7.31}$$

and

$$\delta = (0, 0, \ldots, 1, \ldots, 0) \tag{7.32}$$

in which δ has 1 in the jth position whenever $\alpha_j \geq 1$. Then, by (7.31) and (7.32), we have

$$\gamma \cdot \delta = \alpha_j - 1 \tag{7.33}$$

whenever $\alpha_j \geq 1$. Hence, by (7.30) and (7.33),

$$
\begin{aligned}
f(\xi + \eta) &- \sum_{|\alpha| < N+1} \frac{(\partial^\alpha f)(\xi)}{\alpha!} \eta^\alpha \\
&= \sum_{|\alpha| = N+1} \sum_{\alpha_j \geq 1} \frac{\eta^\alpha}{\alpha!} \alpha_j \int_0^1 (\partial^\alpha f)(\xi + \rho\eta)(1 - \rho)^N d\rho \\
&= (N + 1) \sum_{|\alpha| = N+1} \frac{\eta^\alpha}{\alpha!} \int_0^1 (\partial^\alpha f)(\xi + \rho\eta)(1 - \rho)^N d\rho.
\end{aligned}
\tag{7.34}
$$

Therefore (7.21) is also true for the integer $N + 1$. Hence, by the principle of mathematical induction, the proof of Theorem 7.3 is complete. $\square$

Exercises

7.1. Construct a function ψ in $C_0^\infty(\mathbb{R}^n)$ such that

$$0 \leq \psi(\xi) \leq 1, \quad \xi \in \mathbb{R}^n,$$

$$\psi(\xi) = 1, \quad 1 \leq |\xi| \leq 2,$$

and

$$\psi(\xi) = 0, \quad |\xi| < \frac{1}{2} \text{ or } |\xi| > 4.$$

7.2. For any nonnegative integer N, prove that

$$|z|^{2N} \leq n^N \sum_{|\gamma| = N} |z^\gamma|^2$$

for all $z \in \mathbb{R}^n$.

7.3. Let $f \in C^\infty(\mathbb{R}^n)$ be such that

$$\sup_{x \in \mathbb{R}^n} |(\partial^\alpha f)(x)| < \infty$$

for all multi-indices α. Prove that for every positive integer N, there exists a positive constant C_N such that

$$\left| f(x) - \sum_{|\alpha| < N} \frac{(\partial^\alpha f)(0)}{\alpha!} x^\alpha \right| \le C_N |x|^N, \quad x \in \mathbb{R}^n.$$

Chapter 8

The Product of Two Pseudo-Differential Operators

In this chapter we prove that the product (or composition) of two pseudo-differential operators is again a pseudo-differential operator. We also give an asymptotic expansion for the symbol of the product. The main result is the following theorem.

Theorem 8.1. *Let* $\sigma \in S^{m_1}$ *and* $\tau \in S^{m_2}$. *Then the product* $T_\sigma T_\tau$ *of the pseudo-differential operators* T_σ *and* T_τ *is again a pseudo-differential operator* T_λ, *where* λ *is in* $S^{m_1+m_2}$ *and has the asymptotic expansion*

$$\lambda \sim \sum_\mu \frac{(-i)^{|\mu|}}{\mu!} (\partial_\xi^\mu \sigma)(\partial_x^\mu \tau). \tag{8.1}$$

Here (8.1) means that

$$\lambda - \sum_{|\mu| < N} \frac{(-i)^{|\mu|}}{\mu!} (\partial_\xi^\mu \sigma)(\partial_x^\mu \tau)$$

is a symbol in $S^{m_1+m_2-N}$ *for every positive integer* N.

To motivate the proof, let us begin with an argument to find out how we should proceed. For any function φ in $\mathcal{S}$, we have

$$(T_{\sigma_k}\varphi)(x) = (2\pi)^{-n/2} \int_{\mathbb{R}^n} e^{ix\cdot\xi} \sigma_k(x,\xi) \hat{\varphi}(\xi) \, d\xi$$

for all $x \in \mathbb{R}^n$, where

$$\sigma_k(x,\xi) = \sigma(x,\xi)\varphi_k(\xi)$$

and $\{\varphi_k\}$ is the partition of unity constructed in Theorem 7.1. Hence

$$\sum_{k=0}^\infty (T_{\sigma_k}\varphi)(x) = (2\pi)^{-n/2} \int_{\mathbb{R}^n} e^{ix\cdot\xi} \left\{ \sum_{k=0}^\infty \sigma_k(x,\xi) \right\} \hat{\varphi}(\xi) \, d\xi$$

$$= (2\pi)^{-n/2} \int_{\mathbb{R}^n} e^{ix\cdot\xi} \sigma(x,\xi) \hat{\varphi}(\xi) \, d\xi$$

for all $x \in \mathbb{R}^n$. The interchange of $\sum_{k=0}^{\infty}$ and $\int_{\mathbb{R}^n}$ can of course be justified by Fubini's theorem. Hence

$$T_\sigma \varphi = \sum_{k=0}^{\infty} T_{\sigma_k} \varphi$$

and the convergence of the series can be shown to be absolute and uniform for all $x \in \mathbb{R}^n$. Our goal is to compute the symbol of $T_\sigma T_\tau$. But let us first compute the symbol of $T_{\sigma_k} T_\tau$. You will see why in a minute. Let $\varphi \in \mathcal{S}$. Then, by the definition of a pseudo-differential operator, the Fourier transform, Fubini's theorem and (7.9), we get

$$(T_{\sigma_k} T_\tau \varphi)(x)$$

$$= (2\pi)^{-n/2} \int_{\mathbb{R}^n} e^{ix\cdot\xi} \sigma_k(x, \xi)(T_\tau \varphi)^{\wedge}(\xi)\,d\xi$$

$$= (2\pi)^{-n} \int_{\mathbb{R}^n} e^{ix\cdot\xi} \sigma_k(x, \xi) \left\{ \int_{\mathbb{R}^n} e^{-i\xi\cdot y}(T_\tau \varphi)(y)dy \right\} d\xi$$

$$= (2\pi)^{-n} \int_{\mathbb{R}^n} \left\{ \int_{\mathbb{R}^n} e^{i(x-y)\cdot\xi} \sigma_k(x, \xi)\,d\xi \right\} (T_\tau \varphi)(y)\,dy$$

$$= (2\pi)^{-n/2} \int_{\mathbb{R}^n} K_k(x, x - y)(T_\tau \varphi)(y)\,dy \qquad (8.2)$$

for all $x \in \mathbb{R}^n$. Hence, by the definition of a pseudo-differential operator and Fubini's theorem again, the formula (8.2) becomes

$$\left(T_{\sigma_k} T_\tau \varphi\right)(x)$$

$$= (2\pi)^{-n} \int_{\mathbb{R}^n} K_k(x, x - y) \left\{ \int_{\mathbb{R}^n} e^{iy\cdot\eta} \tau(y, \eta)\hat{\varphi}(\eta)\,d\eta \right\} dy$$

$$= (2\pi)^{-n} \int_{\mathbb{R}^n} e^{ix\cdot\eta} \left\{ \int_{\mathbb{R}^n} e^{-i(x-y)\cdot\eta} K_k(x, x - y)\tau(y, \eta)\,dy \right\} \hat{\varphi}(\eta)\,d\eta$$

$$= (2\pi)^{-n/2} \int_{\mathbb{R}^n} e^{ix\cdot\eta} \lambda_k(x, \eta)\hat{\varphi}(\eta)\,d\eta$$

for all $x \in \mathbb{R}^n$, where

$$\lambda_k(x, \eta) = (2\pi)^{-n/2} \int_{\mathbb{R}^n} e^{-i(x-y)\cdot\eta} K_k(x, x - y)\tau(y, \eta)\,dy.$$

By a simple change of variable,

$$\lambda_k(x, \eta) = (2\pi)^{-n/2} \int_{\mathbb{R}^n} e^{-iz\cdot\eta} K_k(x, z)\tau(x - z, \eta)\,dz \qquad (8.3)$$

for all $x, \eta \in \mathbb{R}^n$. This suggests that

$$(T_\sigma T_\tau \varphi)(x) = (2\pi)^{-n/2} \int_{\mathbb{R}^n} e^{ix\cdot\eta} \lambda(x, \eta)\hat{\varphi}(\eta)\,d\eta$$

for all $x \in \mathbb{R}^n$, where

$$\lambda(x, \eta) = \sum_{k=0}^{\infty} \lambda_k(x, \eta) \qquad (8.4)$$

for all $x, \eta \in \mathbb{R}^n$. Hence all we need to do is to show that $\lambda(x, \eta)$, defined by (8.3) and (8.4), is a symbol in $S^{m_1+m_2}$ and satisfies (8.1).

Remark 8.2. Had we begun with $(T_\sigma T_\tau \varphi)(x)$ instead of $(T_{\sigma_k} T_\tau \varphi)(x)$, we would have the divergent integral $\int_{\mathbb{R}^n} e^{i(x-y)\cdot\xi} \sigma(x, \xi) d\xi$ instead of $\int_{\mathbb{R}^n} e^{i(x-y)\cdot\xi} \sigma_k(x, \xi) d\xi$ in (8.2). This is why we cut off the symbol $\sigma(x, \xi)$ in the ξ variable by the partition of unity $\{\varphi_k\}$.

Proof of Theorem 8.1 For $k = 0, 1, 2, \ldots$, we define λ_k by

$$\lambda_k(x, \xi) = (2\pi)^{-n/2} \int_{\mathbb{R}^n} e^{-iz\cdot\xi} K_k(x, z) \tau(x - z, \xi) \, dz \qquad (8.5)$$

for all $x, \xi \in \mathbb{R}^n$. Now, by the Taylor's formula with integral remainder given in Theorem 7.3, we get

$$\tau(x - z, \xi) = \sum_{|\mu| < N_1} \frac{(-z)^\mu}{\mu!} (\partial_x^\mu \tau)(x, \xi) + R_{N_1}(x, z, \xi), \qquad (8.6)$$

where

$$R_{N_1}(x, z, \xi) = N_1 \sum_{|\mu| = N_1} \frac{(-z)^\mu}{\mu!} \int_0^1 (1 - \theta)^{N_1-1} (\partial_x^\mu \tau)(x - \theta z, \xi) \, d\theta \quad (8.7)$$

for all $x, z, \xi \in \mathbb{R}^n$. Replacing $\tau(x - z, \xi)$ in (8.5) by the right hand side of (8.6), and using (7.8), (7.9), Proposition 4.2 and the Fourier inversion formula, we get

$$\lambda_k(x, \xi) = \sum_{|\mu| < N_1} \frac{(-i)^{|\mu|}}{\mu!} (\partial_\xi^\mu \sigma_k)(x, \xi)(\partial_x^\mu \tau)(x, \xi) + T_{N_1}^{(k)}(x, \xi), \qquad (8.8)$$

where

$$T_{N_1}^{(k)}(x, \xi) = (2\pi)^{-n/2} \int_{\mathbb{R}^n} e^{-iz\cdot\xi} K_k(x, z) R_{N_1}(x, z, \xi) \, dz \qquad (8.9)$$

for all $x, \xi \in \mathbb{R}^n$. For any positive integer N, the function λ given by (8.4) satisfies

$$\lambda - \sum_{|\mu| < N} \frac{(-i)^{|\mu|}}{\mu!} (\partial_\xi^\mu \sigma)(\partial_x^\mu \tau)$$

$$= \lambda - \sum_{|\mu| < N_1} \frac{(-i)^{|\mu|}}{\mu!} (\partial_\xi^\mu \sigma)(\partial_x^\mu \tau) + \sum_{N \le |\mu| < N_1} \frac{(-i)^{|\mu|}}{\mu!} (\partial_\xi^\mu \sigma)(\partial_x^\mu \tau),$$

$$(8.10)$$

where N_1 is any integer larger than N. Obviously,

$$\sum_{N \le |\mu| < N_1} \frac{(-i)^{|\mu|}}{\mu!} (\partial_\xi^\mu \sigma)(\partial_x^\mu \tau) \in S^{m_1 + m_2 - N}.$$

Hence, if we can prove that for all multi-indices α and β, there exists a constant $C_{\alpha,\beta} > 0$ such that

$$\left| \left\{ D_x^\alpha D_\xi^\beta \left[\lambda - \sum_{|\mu| < N_1} \frac{(-i)^{|\mu|}}{\mu!} (\partial_\xi^\mu \sigma)(\partial_x^\mu \tau) \right] \right\} (x,\xi) \right|$$
$$\le C_{\alpha,\beta} (1 + |\xi|)^{m_1 + m_2 - N - |\beta|} \tag{8.11}$$

for all $x, \xi \in \mathbb{R}^n$, then we can conclude that λ is in $S^{m_1 + m_2}$ and has an asymptotic expansion given by (8.1). To this end, we first note that, by (7.8), (8.4) and (8.8),

$$\lambda - \sum_{|\mu| < N_1} \frac{(-i)^{|\mu|}}{\mu!} (\partial_\xi^\mu \sigma)(\partial_x^\mu \tau) = \sum_{k=0}^\infty T_{N_1}^{(k)}. \tag{8.12}$$

Then for any two multi-indices α and β, we need to estimate $D_x^\alpha D_\xi^\beta T_{N_1}^{(k)}$ for all $k = 0, 1, 2, \ldots$. We have the following estimate.

Lemma 8.3. *For all nonnegative integers M, there exists a positive constant C_{α,β,M,N_1} such that*

$$|(D_x^\alpha D_\xi^\beta T_{N_1}^{(k)})(x,\xi)| \le C_{\alpha,\beta,M,N_1} (1 + |\xi|)^{m_2 - 2M} 2^{(m_1 + 2M - N_1)k} \tag{8.13}$$

for all $x, \xi \in \mathbb{R}^n$ and $k = 0, 1, 2, \ldots$.

Let us assume Lemma 8.3 for a moment. Then for all positive integers N, and multi-indices α and β, we can choose a positive integer M such that

$$(1 + |\xi|)^{m_2 - 2M} \le (1 + |\xi|)^{m_1 + m_2 - N - |\beta|} \tag{8.14}$$

for all $\xi \in \mathbb{R}^n$. With this M fixed, we can choose another positive integer N_1 so large that

$$m_1 + 2M - N_1 < 0. \tag{8.15}$$

By (8.12)–(8.15),

$$\left| \left\{ D_x^\alpha D_\xi^\beta \left(\lambda - \sum_{|\mu| < N_1} \frac{(-i)^{|\mu|}}{\mu!} (\partial_\xi^\mu \sigma)(\partial_x^\mu \tau) \right) \right\} (x,\xi) \right|$$
$$\le \sum_{k=0}^\infty C_{\alpha,\beta,M,N_1} (1 + |\xi|)^{m_1 + m_2 - N - |\beta|} 2^{(m_1 + 2M - N_1)k}$$
$$= C_{\alpha,\beta} (1 + |\xi|)^{m_1 + m_2 - N - |\beta|}$$

for all $x, \xi \in \mathbb{R}^n$, where

$$C_{\alpha,\beta} = C_{\alpha,\beta,M,N_1} \sum_{k=0}^{\infty} 2^{(m_1+2M-N_1)k}.$$

$\square$

It remains to prove Lemma 8.3. To this end, we need another lemma.

Lemma 8.4. *Let $R_{N_1}(x, z, \xi)$ be the function given by (8.7). Then for all multi-indices α, β and γ, there exists a constant $C_{\alpha,\beta,\gamma} > 0$ such that*

$$|(\partial_z^\gamma \partial_x^\alpha \partial_\xi^\beta R_{N_1})(x, z, \xi)| \leq C_{\alpha,\beta,\gamma} \left[\sum_{\gamma' \leq \gamma} |z|^{N_1 - |\gamma'|} \right] (1 + |\xi|)^{m_2 - |\beta|}$$

for all $x, z, \xi \in \mathbb{R}^n$.

The proof that Lemma 8.4 implies Lemma 8.3 is by Leibniz's formula, integration by parts, Exercise 4.5 and Theorem 7.2. We leave it as an exercise. See Exercise 8.1.

Proof of Lemma 8.4 By (8.7), we have

$$(\partial_x^\alpha \partial_\xi^\beta R_{N_1})(x, z, \xi)$$

$$= N_1 \sum_{|\mu|=N_1} \frac{(-z)^\mu}{\mu!} \int_0^1 (1-\theta)^{N_1-1} (\partial_x^{\alpha+\mu} \partial_\xi^\beta \tau)(x - \theta z, \xi) \, d\theta \quad (8.16)$$

for all $x, z, \xi \in \mathbb{R}^n$. Hence, by (8.16) and Leibniz's formula,

$$(\partial_z^\gamma \partial_x^\alpha \partial_\xi^\beta R_{N_1})(x, z, \xi)$$

$$= N_1 \sum_{|\mu|=N_1} \sum_{\gamma' \leq \gamma} \binom{\gamma}{\gamma'} \frac{1}{\mu!} \{\partial_z^{\gamma'} (-z)^\mu\} \int_0^1 (1-\theta)^{N_1-1}$$

$$(\partial_x^{\gamma-\gamma'+\alpha+\mu} \partial_\xi^\beta \tau)(x - \theta z, \xi)(-\theta)^{|\gamma-\gamma'|} d\theta \quad (8.17)$$

for all $x, z, \xi \in \mathbb{R}^n$. So, by Exercise 4.5, (8.17) and the fact that $\tau \in S^{m_2}$, there exist positive constants $C_{\gamma'}$ and $C_{\alpha,\beta,\gamma',\mu}$ such that

$$|(\partial_z^\gamma \partial_x^\alpha \partial_\xi^\beta R_{N_1})(x, z, \xi)|$$

$$\leq N_1 \sum_{|\mu|=N_1} \sum_{\gamma' \leq \gamma} \binom{\gamma}{\gamma'} C_{\gamma'} |z|^{N_1 - |\gamma'|} \int_0^1 C_{\alpha,\beta,\gamma',\mu} (1 + |\xi|)^{m_2 - |\beta|} d\theta$$

$$\leq C_{\alpha,\beta,\gamma} \left[\sum_{\gamma' \leq \gamma} |z|^{N_1 - |\gamma'|} \right] (1 + |\xi|)^{m_2 - |\beta|}$$

for all $x, z, \xi \in \mathbb{R}^n$, where

$$C_{\alpha,\beta,\gamma} = N_1 \sum_{|\mu|=N_1} \sup_{\gamma' \leq \gamma} \left\{ \binom{\gamma}{\gamma'} C_{\gamma'} C_{\alpha,\beta,\gamma',\mu} \right\}$$

and this completes the proof of Lemma 8.4. $\square$

Exercises

8.1. Prove that Lemma 8.4 implies Lemma 8.3.

8.2. Let

$$P(x, D) = \sum_{|\alpha| \leq m_1} a_\alpha(x) D^\alpha$$

and

$$Q(x, D) = \sum_{|\alpha| \leq m_2} b_\alpha(x) D^\alpha,$$

where the a_α's and b_α's are in $C^\infty(\mathbb{R}^n)$ and all their partial derivatives are bounded functions on $\mathbb{R}^n$. Compute the symbol of the product $P(x, D)Q(x, D)$ directly. Compare the answer with the symbol obtained by Theorem 8.1.

8.3. Let $q \in C^\infty(\mathbb{R}^n)$ be such that

$$\sup_{x \in \mathbb{R}^n} |(D^\alpha q)(x)| < \infty$$

for all multi-indices α. Let σ be the symbol defined by

$$\sigma(x, \xi) = q(x)$$

for all $x, \xi \in \mathbb{R}^n$. Let τ be any other symbol. Use Theorem 8.1 to compute the symbols of the operators $T_\sigma T_\tau$ and $T_\tau T_\sigma$.

8.4. Let $\sigma \in S^{m_1}$ and $\tau \in S^{m_2}$. Prove that the symbol of the pseudo-differential operator $T_\sigma T_\tau - T_\tau T_\sigma$ is in $S^{m_1+m_2-1}$.

Chapter 9

The Formal Adjoint of a Pseudo-Differential Operator

We begin with a notation. For any pair of functions φ and ψ in $\mathcal{S}$, we define (φ, ψ) by

$$(\varphi, \psi) = \int_{\mathbb{R}^n} \varphi(x)\overline{\psi(x)}\, dx. \tag{9.1}$$

Let σ be a symbol in S^m and T_σ its associated pseudo-differential operator. Suppose there exists a linear operator $T_\sigma^* : \mathcal{S} \to \mathcal{S}$ such that

$$(T_\sigma\varphi, \psi) = (\varphi, T_\sigma^*\psi), \quad \varphi, \psi \in \mathcal{S}. \tag{9.2}$$

Then we call T_σ^* a *formal adjoint* of the operator T_σ. It is very easy to see that a pseudo-differential operator has at most one formal adjoint. Three problems arise.

(1) Does a formal adjoint exist?
(2) If it exists, is it a pseudo-differential operator?
(3) If it is a pseudo-differential operator, can we find an asymptotic expansion for its symbol?

The aim of this chapter is to prove that the formal adjoint of a pseudo-differential operator exists and is a pseudo-differential operator. Moreover, we can obtain a useful asymptotic expansion for the symbol of the formal adjoint. To be more precise, let us prove the following theorem.

Theorem 9.1. *Let σ be a symbol in S^m. Then the formal adjoint of the pseudo-differential operator T_σ is again a pseudo-differential operator T_τ, where τ is a symbol in S^m and has the asymptotic expansion*

$$\tau(x, \xi) \sim \sum_\mu \frac{(-i)^{|\mu|}}{\mu!}(\partial_x^\mu \partial_\xi^\mu \overline{\sigma})(x\,\xi). \tag{9.3}$$

Here (9.3) means that

$$\tau(x,\xi) - \sum_{|\mu|<N} \frac{(-i)^{|\mu|}}{\mu!}(\partial_x^\mu \partial_\xi^\mu \overline{\sigma})(x,\xi)$$

is a symbol in S^{m-N} for every positive integer N.

Before the proof, let us show how the symbol τ can be constructed. For $k = 0, 1, 2, \ldots$, we define σ_k and K_k by (7.8) and (7.9) respectively. Then, by the definition of a formal adjoint,

$$(T_{\sigma_k}\varphi, \psi) = (\varphi, T_{\sigma_k}^*\psi) \tag{9.4}$$

for all φ and ψ in $\mathcal{S}$. By (9.1), Proposition 4.4 (ii), Proposition 4.6 and the definition of a pseudo-differential operator,

$$
\begin{aligned}
&(T_{\sigma_k}\varphi, \psi) \\
&= \int_{\mathbb{R}^n} (T_{\sigma_k}\varphi)(x)\overline{\psi(x)}\, dx \\
&= (2\pi)^{-n/2} \int_{\mathbb{R}^n} \left\{ \int_{\mathbb{R}^n} e^{ix\cdot\eta}\sigma_k(x,\eta)\hat{\varphi}(\eta)\, d\eta \right\} \overline{\psi(x)}\, dx \\
&= (2\pi)^{-n/2} \int_{\mathbb{R}^n} \left\{ \int_{\mathbb{R}^n} \widehat{\sigma_k}(x,y-x)\varphi(y)\, dy \right\} \overline{\psi(x)}\, dx,
\end{aligned}
\tag{9.5}
$$

where

$$\widehat{\sigma_k}(x,y) = (2\pi)^{-n/2} \int_{\mathbb{R}^n} e^{-iy\cdot\eta}\sigma_k(x,\eta)\, d\eta \tag{9.6}$$

for all $x, y \in \mathbb{R}^n$. Therefore, by (7.9), (9.5), (9.6) and Fubini's theorem,

$$
\begin{aligned}
&(T_{\sigma_k}\varphi, \psi) \\
&= (2\pi)^{-n/2} \int_{\mathbb{R}^n} \left\{ \int_{\mathbb{R}^n} \widehat{\sigma_k}(x,y-x)\overline{\psi(x)}\, dx \right\} \varphi(y)\, dy \\
&= (2\pi)^{-n/2} \int_{\mathbb{R}^n} \left\{ \int_{\mathbb{R}^n} K_k(x,x-y)\overline{\psi(x)}\, dx \right\} \varphi(y)\, dy
\end{aligned}
\tag{9.7}
$$

for all $\varphi, \psi \in \mathcal{S}$. Therefore, by (9.4) and (9.7),

$$(T_{\sigma_k}^*\psi)(x) = (2\pi)^{-n/2} \int_{\mathbb{R}^n} \overline{K_k(y,y-x)}\psi(y)\, dy \tag{9.8}$$

for all $x \in \mathbb{R}^n$. Hence, applying the Fourier inversion formula to the function ψ on the right hand side of (9.8), Fubini's theorem and a change of variables, we have

$$(T^*_{\sigma_k}\psi)(x)$$

$$= (2\pi)^{-n} \int_{\mathbb{R}^n} \overline{K_k}(y, y-x) \left\{ \int_{\mathbb{R}^n} e^{iy\cdot\eta}\hat{\psi}(\eta)\,d\eta \right\} dy$$

$$= (2\pi)^{-n} \int_{\mathbb{R}^n} \left\{ \int_{\mathbb{R}^n} e^{iy\cdot\eta}\overline{K_k}(y, y-x)\,dy \right\} \hat{\psi}(\eta)\,d\eta$$

$$= (2\pi)^{-n} \int_{\mathbb{R}^n} e^{ix\cdot\eta} \left\{ \int_{\mathbb{R}^n} e^{i\eta\cdot(y-x)}\overline{K_k}(y, y-x)\,dy \right\} \hat{\psi}(\eta)\,d\eta$$

$$= (2\pi)^{-n} \int_{\mathbb{R}^n} e^{ix\cdot\eta} \left\{ \int_{\mathbb{R}^n} e^{i\eta\cdot z}\overline{K_k}(x+z, z)\,dz \right\} \hat{\psi}(\eta)\,d\eta \tag{9.9}$$

for all $x \in \mathbb{R}^n$. It is clear from (9.9) that

$$T^*_{\sigma_k} = T_{\tau_k}, \tag{9.10}$$

where

$$\tau_k(x, \eta) = (2\pi)^{-n/2} \int_{\mathbb{R}^n} e^{i\eta\cdot z}\overline{K_k}(x+z, z)\,dz. \tag{9.11}$$

Since

$$(T_\sigma\,\varphi, \psi) = \sum_{k=0}^{\infty}(T_{\sigma_k}\varphi, \psi)$$

for all $\varphi, \psi \in \mathcal{S}$, it is clear from (9.4) and (9.10) that a good candidate for τ is given by

$$\tau(x, \eta) = \sum_{k=0}^{\infty} \tau_k(x, \eta) \tag{9.12}$$

for all $x, \eta \in \mathbb{R}^n$. Hence it remains to prove that τ is a symbol of order m with an asymptotic expansion given by (9.3) and

$$(T_\sigma\varphi, \psi) = (\varphi, T_\tau\psi) \tag{9.13}$$

for all $\varphi, \psi \in \mathcal{S}$.

Proof of Theorem 9.1 For $k = 0, 1, 2, \ldots$, define τ_k by (9.11). Let N_1 be any positive integer. Then, by the Taylor's formula with integral remainder given in Theorem 7.3,

$$K_k(x+z, z) = \sum_{|\mu| < N_1} \frac{z^\mu}{\mu!}(\partial_x^\mu K_k)(x, z) + R_{N_1}^{(k)}(x, z), \tag{9.14}$$

where, by (7.9),

$$R_{N_1}^{(k)}(x, z)$$

$$= N_1 \sum_{|\mu|=N_1} \frac{z^\mu}{\mu!} \int_0^1 (1-\theta)^{N_1-1} (\partial_x^\mu K_k)(x+\theta z, z)\, d\theta$$

$$= N_1 \sum_{|\mu|=N_1} \frac{z^\mu}{\mu!} \int_0^1 (1-\theta)^{N_1-1} (2\pi)^{-n/2} \int_{\mathbb{R}^n} e^{iz\cdot\xi} (\partial_x^\mu \sigma_k)(x+\theta z, \xi)\, d\xi\, d\theta.$$

$$(9.15)$$

Hence, by (7.9), (9.11), (9.14) and Proposition 4.2,

$$\tau_k(x, \eta) = \sum_{|\mu|<N_1} \frac{(-i)^{|\mu|}}{\mu!} (\partial_x^\mu \partial_\eta^\mu \overline{\sigma_k})(x, \eta) + T_{N_1}^{(k)}(x, \eta), \qquad (9.16)$$

where

$$T_{N_1}^{(k)}(x, \eta) = (2\pi)^{-n/2} \int_{\mathbb{R}^n} e^{i\eta\cdot z} \overline{R_{N_1}^{(k)}}(x, z)\, dz \qquad (9.17)$$

for all $x, \eta \in \mathbb{R}^n$. For any positive integer N,

$$\tau - \sum_{|\mu|<N} \frac{(-i)^{|\mu|}}{\mu!} \partial_x^\mu \partial_\eta^\mu \overline{\sigma}$$

$$= \tau - \sum_{|\mu|<N_1} \frac{(-i)^{|\mu|}}{\mu!} \partial_x^\mu \partial_\eta^\mu \overline{\sigma} + \sum_{N\le|\mu|<N_1} \frac{(-i)^{|\mu|}}{\mu!} \partial_x^\mu \partial_\eta^\mu \overline{\sigma}, \qquad (9.18)$$

where N_1 is any integer larger than N. Obviously,

$$\sum_{N\le|\mu|<N_1} \frac{(-i)^{|\mu|}}{\mu!} \partial_x^\mu \partial_\eta^\mu \overline{\sigma} \in S^{m-N}.$$

Hence, if we can prove that for all multi-indices α and β, there exists a constant $C_{\alpha,\beta} > 0$ such that

$$\left| \left[D_x^\alpha D_\eta^\beta \left\{ \tau - \sum_{|\mu|<N_1} \frac{(-i)^{|\mu|}}{\mu!} \partial_x^\mu \partial_\eta^\mu \overline{\sigma} \right\} \right](x, \eta) \right|$$

$$\le C_{\alpha,\beta} (1+|\eta|)^{m-N-|\beta|} \qquad (9.19)$$

for all $x, \eta \in \mathbb{R}^n$, then we can conclude that $\tau \in S^m$ and has an asymptotic expansion given by (9.3). To this end, we first note that, by (7.8), (9.12) and (9.16),

$$\tau - \sum_{|\mu|<N_1} \frac{(-i)^{|\mu|}}{\mu!} \partial_x^\mu \partial_\eta^\mu \overline{\sigma} = \sum_{k=0}^\infty T_{N_1}^{(k)}. \qquad (9.20)$$

Let α and β be any two multi-indices. Then, by (9.17) and an integration by parts,

$$(D_x^\alpha D_\eta^\beta T_{N_1}^{(k)})(x,\eta)$$

$$= (2\pi)^{-n/2} \int_{\mathbb{R}^n} e^{i\eta \cdot z} z^\beta \left(D_x^\alpha \overline{R_{N_1}^{(k)}} \right)(x,z)\, dz$$

$$= (1+|\eta|^2)^{-K} (2\pi)^{-n/2} \int_{\mathbb{R}^n} e^{i\eta \cdot z} (1-\Delta_z)^K \left\{ z^\beta \left(D_x^\alpha \overline{R_{N_1}^{(k)}} \right)(x,z) \right\} dz,$$

$$(9.21)$$

where K is any positive integer. Let $P(D) = (1-\Delta)^K$. Then, by (9.15), Leibniz's formulas and an integration by parts,

$$(1-\Delta_z)^K \left\{ z^\beta \left(D_x^\alpha \overline{R_{N_1}^{(k)}} \right)(x,z) \right\}$$

$$= (1-\Delta_z)^K N_1 \sum_{|\mu|=N_1} \frac{z^{\mu+\beta}}{\mu!} \int_0^1 (1-\theta)^{N_1-1} (2\pi)^{-n/2}$$

$$\int_{\mathbb{R}^n} (-i)^{|\alpha|} e^{-iz \cdot \xi} (\partial_x^{\alpha+\mu} \overline{\sigma_k})(x+\theta z, \xi)\, d\xi\, d\theta$$

$$= N_1 \sum_{|\mu|=N_1} \sum_{|\delta| \le 2K} \frac{P^{(\delta)}(D)(z^{\mu+\beta})}{\mu!\delta!} \int_0^1 (1-\theta)^{N_1-1} (2\pi)^{-n/2}$$

$$\int_{\mathbb{R}^n} (\cdots)\, d\xi\, d\theta,$$

$$(9.22)$$

where

$$(\cdots) = (-i)^{|\alpha|} \sum_{\rho \le \delta} \binom{\delta}{\rho} \theta^{|\delta-\rho|} (D_z^\rho e^{-iz\cdot\xi})(D_x^{\delta-\rho} \partial_x^{\alpha+\mu} \overline{\sigma_k})(x+\theta z, \xi). \quad (9.23)$$

Let γ be any multi-index. Then, by (9.22), (9.23), an integration by parts and Leibniz's formula,

$$(1-\Delta_z)^K \left\{ z^\beta \left(D_x^\alpha \overline{R_{N_1}^{(k)}} \right)(x,z) \right\}$$

$$= N_1 \sum_{|\mu|=N_1} \sum_{|\delta| \le 2K} \sum_{\rho \le \delta} \binom{\delta}{\rho} \frac{P^{(\delta)}(D)(z^{\mu+\beta})}{\mu!\delta!} \int_0^1 (1-\theta)^{N_1-1}$$

$$\theta^{|\delta-\rho|} (2\pi)^{-n/2} (\ast\ast\ast)\, d\theta,$$

$$(9.24)$$

where $z^\gamma(\ast\ast\ast)$ is equal to

$$\int_{\mathbb{R}^n} e^{-iz\cdot\xi} (-i)^{|\alpha|} (-1)^{|\rho|} \sum_{\gamma' \le \gamma} \binom{\gamma}{\gamma'}$$

$$\{ D_\xi^{\gamma'} (\xi^\rho D_x^{\delta-\rho} \partial_x^{\alpha+\mu} \overline{\sigma}) \}(x+\theta z, \xi)(D^{\gamma-\gamma'} \varphi_k)(\xi)\, d\xi. \quad (9.25)$$

Using the fact that σ is a symbol in S^m, (9.25) and Theorem 7.1 (vi), we can find a constant $C > 0$, depending on $\gamma', \rho, \delta, \alpha, \mu$ but not on k, such that

$$|z^\gamma(***)| \leq \int_{W_k} \sum_{\gamma' \leq \gamma} C(1 + |\xi|)^{m+|\rho|-|\gamma'|} 2^{-k|\gamma-\gamma'|} \, d\xi, \tag{9.26}$$

where W_0 and W_k, $k = 1, 2, \ldots$, are given by (7.12) and (7.13) respectively. Let M be any positive integer. Then, by (9.26) and Exercise 7.2, there is another constant $C > 0$, depending on $M, \rho, \delta, \alpha, \mu$ but not on k, such that

$$|(***)| \leq C|z|^{-2M} 2^{k(m+|\rho|-2M+n)}. \tag{9.27}$$

Hence, by (9.24), (9.27) and Exercise 9.5, there is a constant $C > 0$, depending on α, β, K, M, N_1 but not on k, such that

$$\left| (1 - \Delta_z)^K \left\{ z^\beta \left(D_x^\alpha \overline{R_{N_1}^{(k)}} \right)(x, z) \right\} \right|$$
$$\leq C|z|^{-2M} \left\{ \sum_{|\delta| \leq 2K} |z|^{|\beta|+N_1-|\delta|} \right\} 2^{k(m+2K-2M+n)}. \tag{9.28}$$

We choose K so large that

$$(1 + |\eta|)^{-2K} \leq (1 + |\eta|)^{m-N-|\beta|} \tag{9.29}$$

for all $\eta \in \mathbb{R}^n$. Then we choose M so large, say, equal to M', that

$$m + 2K - 2M' + n < 0. \tag{9.30}$$

Then we choose N_1 so large that

$$\int_{|z| \leq 1} |z|^{-2M'} \left\{ \sum_{|\delta| \leq 2K} |z|^{|\beta|+N_1-|\delta|} \right\} dz < \infty. \tag{9.31}$$

Hence, by (9.28) and (9.31), there exists a constant $C_1 > 0$, depending on $\alpha, \beta, K, M', N_1$ but not on k, such that

$$\int_{|z| \leq 1} \left| (1 - \Delta_z)^K \left\{ z^\beta \left(D_x^\alpha \overline{R_{N_1}^{(k)}} \right)(x, z) \right\} \right| dz \leq C_1 2^{k(m+2K-2M'+n)}. \tag{9.32}$$

Now, we choose M so large, say, equal to M'', that

$$m + 2K - 2M'' + n < 0 \tag{9.33}$$

and

$$\int_{|z| \geq 1} |z|^{-2M''} \left\{ \sum_{|\delta| \leq 2K} |z|^{|\beta|+N_1-|\delta|} \right\} dz < \infty. \tag{9.34}$$

Hence, by (9.28) and (9.34), there exists a constant $C_2 > 0$, depending on $\alpha, \beta, K, M'', N_1$ but not on k, such that

$$\int_{|z| \geq 1} \left| (1 - \Delta_z)^K \left\{ z^\beta \left(D_x^\alpha \overline{R_{N_1}^{(k)}} \right) (x, z) \right\} \right| dz$$

$$\leq C_2 \, 2^{k(m+2K-2M''+n)}. \tag{9.35}$$

Hence, by (9.21), (9.32) and (9.35), we get another constant $C > 0$, depending on $\alpha, \beta, K, M', M'', N_1$ but not on k, such that

$$|(D_x^\alpha D_\eta^\beta T_{N_1}^{(k)})(x, \eta)|$$

$$\leq C(1 + |\eta|^2)^{-K} \{ 2^{k(m+2K-2M'+n)} + 2^{k(m+2K-2M''+n)} \}. \tag{9.36}$$

Hence, for any two multi-indices α and β, we have, by (9.20), (9.29), (9.30), (9.33) and (9.36), a constant $C_{\alpha,\beta} > 0$ for which (9.19) is valid for all $x, \eta \in \mathbb{R}^n$. Therefore the function τ defined by (9.12) is a symbol in S^m and has an asymptotic expansion given by (9.3). That (9.13) is true should be by now obvious. At any rate, it is a simple consequence of Theorem 7.1 (ii), (7.8), (9.4), (9.10) and (9.12). $\qquad\square$

Exercises

9.1. Prove that a pseudo-differential operator has a unique formal adjoint.

9.2. Let σ and τ be any two symbols. Prove that

$$(T_\sigma^*)^* = T_\sigma$$

and

$$(T_\sigma T_\tau)^* = T_\tau^* T_\sigma^*.$$

9.3. Let $P(x, D) = \sum_{|\alpha| \leq m} a_\alpha(x) D^\alpha$, where the a_α's are in $C^\infty(\mathbb{R}^n)$ and all their partial derivatives are bounded functions on $\mathbb{R}^n$. Compute the symbol of the formal adjoint of $P(x, D)$ directly. Compare the answer with the symbol obtained by Theorem 9.1.

9.4. Let σ and τ be as in Exercise 8.3. Use Theorem 9.1 to compute the symbols of the formal adjoints of $T_\sigma + T_\tau$, $T_\sigma T_\tau$ and $T_\tau T_\sigma$.

9.5. In deriving (9.28) from (9.24) and (9.27), we use the fact that there exists a positive constant C, depending on μ, β and K only, such that

$$\sum_{|\delta| \leq 2K} |P^{(\delta)}(D)(z^{\mu+\beta})| \leq C \sum_{|\delta| \leq 2K} |z|^{|\mu|+|\beta|-|\delta|}$$

for all $z \in \mathbb{R}^n$. Prove the fact.

Chapter 10

The Parametrix of an Elliptic Pseudo-Differential Operator

Among all pseudo-differential operators there exists a class of operators which come up frequently in applications and are particularly easy to work with. They are called elliptic operators. They are nice because they have approximate inverses (or parametrices) which are also pseudo-differential operators. Our first task is to make these concepts precise.

A symbol σ in S^m is said to be *elliptic* if there exist positive constants C and R such that

$$|\sigma(x,\xi)| \geq C(1+|\xi|)^m, \quad |\xi| \geq R.$$

Of course, a pseudo-differential operator T_σ is said to be *elliptic* if its symbol is elliptic.

Theorem 10.1. *Let σ be an elliptic symbol in S^m. Then there exists a symbol τ in S^{-m} such that*

$$T_\tau T_\sigma = I + R \tag{10.1}$$

and

$$T_\sigma T_\tau = I + S, \tag{10.2}$$

where R and S are pseudo-differential operators with symbols in $\cap_{k \in \mathbb{R}} S^k$, and I is the identity operator.

Remark 10.2. In other words, Theorem 10.1 says that if T_σ is an elliptic pseudo-differential operator, then it can be inverted modulo some error terms R and S with symbols in $\cap_{k \in \mathbb{R}} S^k$. In the theory of regularity of solutions of partial differential equations, operators with symbols in $\cap_{k \in \mathbb{R}} S^k$ are called *infinitely smoothing* and can be neglected. (See Chapter 15 for a discussion of infinitely smoothing operators and regularity theory.) For these reasons, we call T_τ an approximate inverse, or more often in the literature, a *parametrix* of T_σ.

We first give a proof of (10.1). The idea is to find a sequence of symbols $\tau_j \in S^{-m-j}$, $j = 0, 1, 2, \ldots$. Let us assume that this has been done. Then, by Theorem 6.10, there exists a symbol $\tau \in S^{-m}$ such that $\tau \sim \sum_{j=0}^{\infty} \tau_j$. Then, by the product formula given in Theorem 8.1, the symbol λ of the product $T_\tau T_\sigma$ is in S^0 and such that

$$\lambda - \sum_{|\gamma| < N} \frac{(-i)^{|\gamma|}}{\gamma!} (\partial_x^\gamma \sigma)(\partial_\xi^\gamma \tau) \in S^{-N} \tag{10.3}$$

for every positive integer N. But $\tau \sim \sum_{j=0}^{\infty} \tau_j$ implies that

$$\tau - \sum_{j=0}^{N-1} \tau_j \in S^{-m-N} \tag{10.4}$$

for every positive integer N. Hence, by (10.3) and (10.4),

$$\lambda - \sum_{|\gamma| < N} \frac{(-i)^{|\gamma|}}{\gamma!} (\partial_x^\gamma \sigma) \sum_{j=0}^{N-1} (\partial_\xi^\gamma \tau_j) \in S^{-N} \tag{10.5}$$

for every positive integer N. But we can write

$$\sum_{|\gamma| < N} \frac{(-i)^{|\gamma|}}{\gamma!} \sum_{j=0}^{N-1} (\partial_\xi^\gamma \tau_j)(\partial_x^\gamma \sigma)$$

$$= \tau_0 \sigma + \sum_{l=1}^{N-1} \left\{ \tau_l \sigma + \sum_{\substack{|\gamma|+j=l \\ j<l}} \frac{(-i)^{|\gamma|}}{\gamma!} (\partial_\xi^\gamma \tau_j)(\partial_x^\gamma \sigma) \right\}$$

$$+ \sum_{\substack{\cdot \, |\gamma|+j \geq N \\ |\gamma| < N, \, j < l}} \frac{(-i)^{|\gamma|}}{\gamma!} (\partial_\xi^\gamma \tau_j)(\partial_x^\gamma \sigma). \tag{10.6}$$

To simplify (10.6), we choose τ_j, $j = 0, 1, 2, \ldots$, in the following way. Define τ_0 by

$$\tau_0(x, \xi) = \begin{cases} \dfrac{\psi(\xi)}{\sigma(x,\xi)}, & |\xi| > R, \\ 0, & |\xi| \leq R, \end{cases} \tag{10.7}$$

where ψ is any function in $C^\infty(\mathbb{R}^n)$ such that $\psi(\xi) = 1$ for $|\xi| \geq 2R$ and $\psi(\xi) = 0$ for $|\xi| \leq R$, and we define τ_l for $l \geq 1$ inductively by

$$\tau_l = - \left\{ \sum_{\substack{|\gamma|+j=l \\ j<l}} \frac{(-i)^{|\gamma|}}{\gamma!} (\partial_x^\gamma \sigma)(\partial_\xi^\gamma \tau_j) \right\} \tau_0. \tag{10.8}$$

Then it can be checked easily that $\tau_j \in S^{-m-j}$, $j = 0, 1, 2, \ldots$. (See Exercise 10.7.) Now, by (10.7), $\tau_0 \sigma = 1$ for $|\xi| \geq 2R$. The second term on the right hand side of (10.6) vanishes for $|\xi| \geq 2R$ by (10.7) and (10.8). As for the third term there, we see easily that

$$(\partial_\xi^\gamma \tau_j)(\partial_x^\gamma \sigma) \in S^{-N}$$

whenever $|\gamma| + j \geq N$. Hence, by (10.6),

$$\sum_{|\gamma| < N} \frac{(-i)^{|\gamma|}}{\gamma!} \sum_{j=0}^{N-1} (\partial_\xi^\gamma \tau_j)(\partial_x^\gamma \sigma) - 1 \in S^{-N} \tag{10.9}$$

for every positive integer N. Thus, by (10.5) and (10.9),

$$\lambda - 1 \in S^{-N}$$

for every positive integer N. Hence, if we pick R to be the pseudo-differential operator with symbol $\lambda - 1$, then the proof of (10.1) is complete.

By a similar argument, we can find another symbol κ in S^{-m} such that

$$T_\sigma T_\kappa = I + R', \tag{10.10}$$

where R' is a pseudo-differential operator with symbol in $\cap_{k \in \mathbb{R}} S^k$. (See Exercise 10.8.) By (10.1) and (10.10),

$$T_\kappa + RT_\kappa = T_\tau + T_\tau R'.$$

Since RT_κ and $T_\tau R'$ are pseudo-differential operators with symbols in $\cap_{k \in \mathbb{R}} S^k$, it follows that

$$T_\kappa = T_\tau + R'', \tag{10.11}$$

where

$$R'' = T_\tau R' - RT_\kappa$$

is another pseudo-differential operator with symbol in $\cap_{k \in \mathbb{R}} S^k$. Hence, by (10.10) and (10.11),

$$T_\sigma T_\tau = I + S,$$

where

$$S = R' - T_\sigma R''.$$

Since S is a pseudo-differential operator with symbol in $\cap_{k \in \mathbb{R}} S^k$, it follows that (10.2) is proved.

The following theorem tells us that only elliptic pseudo-differential operators have parametrices.

Theorem 10.3. *Let $\sigma \in S^m$ be such that there exists a $\tau \in S^{-m}$ for which (10.1) or (10.2) is true, where R and S are infinitely smoothing pseudo-differential operators, and I is the identity operator. Then σ is elliptic.*

Remark 10.4. If (10.1) (or (10.2)) is true, then we call T_τ a left (or right) parametrix of T_σ. Thus, a consequence of Theorems 10.1 and 10.3 is that if a pseudo-differential operator T_σ has a left (or right) parametrix T_τ, then T_τ is also a right (or left) parametrix of T_σ.

Proof of Theorem 10.3 Let us first assume that (10.1) is valid and let r be the symbol of R. Then, by Theorem 8.1,

$$1 + r - \sigma\tau = \delta, \tag{10.12}$$

where δ is some symbol in S^{-1}. Since $\tau \in S^{-m}$, we can find a positive constant C_1 such that

$$|\tau(x,\xi)| \leq C_1(1 + |\xi|)^{-m}, \quad x, \xi \in \mathbb{R}^n. \tag{10.13}$$

Thus, by (10.12) and (10.13),

$$|1 + r(x,\xi) - \delta(x,\xi)| \leq C_1|\sigma(x,\xi)|(1 + |\xi|)^{-m}, \quad x, \xi \in \mathbb{R}^n,$$

and hence

$$|\sigma(x,\xi)| \geq \frac{1}{C_1}(1 + |\xi|)^m(1 - |\delta(x,\xi)| - |r(x,\xi)|), \quad x, \xi \in \mathbb{R}^n. \tag{10.14}$$

Since $\delta \in S^{-1}$, it follows that there is a positive constant C_2 such that

$$|\delta(x,\xi)| \leq C_2(1 + |\xi|)^{-1}, \quad x, \xi \in \mathbb{R}^n. \tag{10.15}$$

Since $r \in \cap_{k \in \mathbb{R}}S^k$, we can find a positive constant C_3 such that

$$|r(x,\xi)| \leq C_3(1 + |\xi|)^{-1}, \quad x, \xi \in \mathbb{R}^n. \tag{10.16}$$

So, by (10.14)–(10.16),

$$|\sigma(x,\xi)| \geq \frac{1}{C_1}(1 + |\xi|)^m\{1 - (C_2 + C_3)(1 + |\xi|)^{-1}\}, \quad x, \xi \in \mathbb{R}^n. \tag{10.17}$$

Now, let R be any positive number such that

$$1 - (C_2 + C_3)(1 + |\xi|)^{-1} \geq \frac{1}{2}, \quad |\xi| > R. \tag{10.18}$$

Then, by (10.17) and (10.18), we get

$$|\sigma(x,\xi)| \geq \frac{1}{2C_1}(1 + |\xi|)^m, \quad |\xi| > R,$$

and this completes the proof under the assumption that (10.1) is valid. The proof for the case when (10.2) is valid is similar and hence left as an exercise. See Exercise 10.9.   □

Exercises

10.1. (i) Let $P(x, D) = \sum_{|\alpha| \leq m} a_\alpha(x) D^\alpha$, where the a_α's are in $C^\infty(\mathbb{R}^n)$ and all their partial derivatives are bounded functions on $\mathbb{R}^n$. We call $\sum_{|\alpha|=m} a_\alpha(x)\xi^\alpha$ the *principal symbol* of $P(x, D)$ and denote it by $P_m(x, \xi)$. Prove that $P(x, D)$ is elliptic if and only if there exist positive constants C and R such that

$$|P_m(x, \xi)| \geq C(1 + |\xi|)^m, \quad |\xi| \geq R.$$

(ii) Let $P(D) = \sum_{|\alpha| \leq m} a_\alpha D^\alpha$ be a linear partial differential operator with constant coefficients. Let $P_m(\xi)$ be the principal symbol of $P(D)$. Prove that $P(D)$ is elliptic if and only if

$$P_m(\xi) = 0, \, \xi \in \mathbb{R}^n \Rightarrow \xi = 0.$$

10.2. Let T_σ and T_τ be elliptic pseudo-differential operators. Prove that the product $T_\sigma T_\tau$ is also elliptic.

10.3. Let T_σ be an elliptic pseudo-differential operator. Prove that the formal adjoint T_σ^* of T_σ is also elliptic.

10.4. Prove that any two parametrices of an elliptic pseudo-differential operator differ by an infinitely smoothing operator.

10.5. Is a parametrix of an elliptic pseudo-differential operator elliptic? Explain your answer.

10.6. Prove that, in the proof of Theorem 10.1, (10.3) and (10.4) imply (10.5).

10.7. Let $\{\tau_j\}_{j=0}^\infty$ be the sequence of functions defined by (10.7) and (10.8). Prove that τ_j is a symbol in S^{-m-j} for $j = 0, 1, 2, \ldots$.

10.8. Prove that there exists a symbol κ in S^{-m} such that (10.10) is valid.

10.9. Let $\sigma \in S^m$ be such that there exists a $\tau \in S^{-m}$ for which (10.2) is true, where S is an infinitely smoothing operator and I is the identity operator. Prove that σ is elliptic.

10.10. Find a parametrix of an elliptic linear partial differential operator with constant coefficients on $\mathbb{R}^n$.

Chapter 11

L^p-Boundedness of Pseudo-Differential Operators

Let σ be a symbol. Then, by Proposition 6.7, the pseudo-differential operator T_σ maps the Schwartz space $\mathcal{S}$ into $\mathcal{S}$. In fact, the following proposition is true.

Proposition 11.1. T_σ maps $\mathcal{S}$ continuously into $\mathcal{S}$. More precisely, if $\varphi_k \to 0$ in $\mathcal{S}$, then $T_\sigma \varphi_k \to 0$ in $\mathcal{S}$ as $k \to \infty$.

To prove Proposition 11.1, we need some preliminary results.

Lemma 11.2. If $\varphi_k \to 0$ in $\mathcal{S}$ as $k \to \infty$, then $\varphi_k \to 0$ in $L^p(\mathbb{R}^n)$ as $k \to \infty$ for $1 \le p \le \infty$.

Proof If $\varphi_k \to 0$ in $\mathcal{S}$ as $k \to \infty$, then $\varphi_k \to 0$ uniformly on $\mathbb{R}^n$ as $k \to \infty$. This proves the lemma for $p = \infty$. So, consider $1 \le p < \infty$. Then for any positive integer N with $N > \frac{n}{p}$, we have

$$(1 + |x|)^N |\varphi_k(x)| \to 0$$

uniformly on $\mathbb{R}^n$ as $k \to \infty$. Hence for k large enough,

$$|\varphi_k(x)| \le (1 + |x|)^{-N}, \quad x \in \mathbb{R}^n.$$

Since $(1 + |x|)^{-Np} \in L^1(\mathbb{R}^n)$, it follows from Lebesgue's dominated convergence theorem that

$$\int_{\mathbb{R}^n} |\varphi_k(x)|^p dx \to 0$$

as $k \to \infty$. This completes the proof. $\qquad \square$

Lemma 11.3. The Fourier transformation $\mathcal{F}$ maps $\mathcal{S}$ continuously into $\mathcal{S}$. More precisely, if $\varphi_k \to 0$ in $\mathcal{S}$ as $k \to \infty$, then $\widehat{\varphi_k} \to 0$ in $\mathcal{S}$ as $k \to \infty$.

75

Proof Let α and β be any two multi-indices. Then, by Proposition 4.2,

$$|\xi^\alpha (D^\beta \widehat{\varphi_k})(\xi)| = |\{D^\alpha((-x)^\beta \varphi_k)\}^\wedge(\xi)|$$
$$\leq (2\pi)^{-n/2}\|D^\alpha((-x)^\beta \varphi_k)\|_1, \quad \xi \in \mathbb{R}^n. \qquad (11.1)$$

Since $\varphi_k \to 0$ in $\mathcal{S}$, it follows that $D^\alpha((-x)^\beta \varphi_k) \to 0$ in $\mathcal{S}$ as $k \to \infty$. By Lemma 11.2, $\|D^\alpha((-x)^\beta \varphi_k)\|_1 \to 0$ as $k \to \infty$. By (11.1),

$$\sup_{\xi \in \mathbb{R}^n} |\xi^\alpha (D^\beta \widehat{\varphi_k})(\xi)| \to 0$$

as $k \to \infty$. This proves that $\widehat{\varphi_k} \to 0$ in $\mathcal{S}$ as $k \to \infty$. $\qquad \square$

Proof of Proposition 11.1 Suppose $\sigma \in S^m$. Then for any two multi-indices α and β, we have, by (6.12), positive constants $C_{\alpha,\beta,\gamma,\delta}$, depending on α, β, γ and δ only, such that

$$\sup_{x \in \mathbb{R}^n} |x^\alpha (D^\beta (T_\sigma \varphi_k))(x)|$$
$$\leq (2\pi)^{-n/2} \sum_{\gamma \leq \beta} \sum_{\delta \leq \alpha} \binom{\beta}{\gamma}\binom{\alpha}{\delta} C_{\alpha,\beta,\gamma,\delta} \int_{\mathbb{R}^n} (1 + |\xi|)^{m-|\alpha|+|\delta|}$$
$$\left|D_\xi^\delta (\xi^\gamma \hat{\varphi}_k(\xi))\right| d\xi. \qquad (11.2)$$

Since $\varphi_k \to 0$ in $\mathcal{S}$ as $k \to \infty$, it follows easily from Lemma 11.3 that

$$(1 + |\xi|^2)^{(m-|\alpha|+|\delta|)/2} D_\xi^\delta(\xi^\gamma \hat{\varphi}_k(\xi)) \to 0$$

in $\mathcal{S}$ as $k \to \infty$. Hence, by Lemma 11.2, the integral on the right hand side of (11.2) goes to zero as $k \to \infty$. This proves that $T_\sigma \varphi_k \to 0$ in $\mathcal{S}$ as $k \to \infty$. $\qquad \square$

The pseudo-differential operator T_σ, initially defined on the Schwartz space $\mathcal{S}$, can be extended to a linear mapping defined on the space $\mathcal{S}'$ of tempered distributions. To wit, take a distribution u in $\mathcal{S}'$ and define $T_\sigma u$ by

$$(T_\sigma u)(\varphi) = u(\overline{T_\sigma^* \overline{\varphi}}), \quad \varphi \in \mathcal{S}, \qquad (11.3)$$

where T_σ^* is the formal adjoint of T_σ introduced in Chapter 9.

Proposition 11.4. *T_σ is a linear mapping from $\mathcal{S}'$ into $\mathcal{S}'$.*

Proof Let $u \in \mathcal{S}'$. Then for any sequence $\{\varphi_k\}$ of functions in $\mathcal{S}$ converging to zero in $\mathcal{S}$, we have, by (11.3),

$$(T_\sigma u)(\varphi_k) = u(\overline{T_\sigma^* \overline{\varphi_k}}), \quad k = 1, 2, \ldots. \qquad (11.4)$$

By Proposition 11.1, $T_\sigma^* \overline{\varphi_k} \to 0$ in $\mathcal{S}$ as $k \to \infty$. Hence, using (11.4) and the fact that u is a tempered distribution, we conclude that $(T_\sigma u)(\varphi_k) \to 0$ as $k \to \infty$. Hence $T_\sigma u \in \mathcal{S}'$. $\qquad\square$

To enquire whether T_σ maps $\mathcal{S}'$ continuously into $\mathcal{S}'$ or not, we need a notion of convergence in $\mathcal{S}'$.

Definition 11.5. A sequence of distributions $\{u_k\}$ in $\mathcal{S}'$ is said to converge to zero in $\mathcal{S}'$ (denoted by $u_k \to 0$ in $\mathcal{S}'$) if $u_k(\varphi) \to 0$ as $k \to \infty$ for all $\varphi \in \mathcal{S}$.

Proposition 11.6. T_σ maps $\mathcal{S}'$ continuously into $\mathcal{S}'$. More precisely, if $u_k \to 0$ in $\mathcal{S}'$ as $k \to \infty$, then $T_\sigma u_k \to 0$ in $\mathcal{S}'$ as $k \to \infty$.

Proof We need only check Definition 11.5 for the sequence $\{T_\sigma u_k\}$. To do this, let $\varphi \in \mathcal{S}$. Then, using (11.3) and the fact that $u_k \to 0$ in $\mathcal{S}'$ as $k \to \infty$,

$$(T_\sigma u_k)(\varphi) = u_k(\overline{T_\sigma^* \overline{\varphi}}) \to 0$$

as $k \to \infty$. Hence $T_\sigma u_k \to 0$ in $\mathcal{S}'$ as $k \to \infty$, and the proof is complete.$\square$

Let us recall that, by Proposition 5.5 and Remark 5.6, every function f in $L^p(\mathbb{R}^n)$ is a tempered distribution. Hence, by Proposition 11.4, $T_\sigma f$ is also a tempered distribution. What sort of a tempered distribution is it? We answer this question in the following theorem and Theorem 12.9 in Chapter 12.

Theorem 11.7. Let σ be a symbol in S^0. Then $T_\sigma : L^p(\mathbb{R}^n) \to L^p(\mathbb{R}^n)$ is a bounded linear operator for $1 < p < \infty$.

The following result plays an important role in our proof of Theorem 11.7. It is a special case of Theorem 2.5 in [Hörmander (1960)]. Its proof is outside the scope of this book and hence omitted.

Theorem 11.8. Let $m \in C^k(\mathbb{R}^n - \{0\})$, $k > \frac{n}{2}$, be such that there is a positive constant B for which

$$|(D^\alpha m)(\xi)| \leq B|\xi|^{-|\alpha|}, \quad \xi \neq 0,$$

for all multi-indices α with $|\alpha| \leq k$. Then for $1 < p < \infty$, there is a positive constant C, depending on p and n only, such that

$$\|T\varphi\|_p \leq CB\|\varphi\|_p, \quad \varphi \in \mathcal{S},$$

where

$$(T\varphi)(x) = (2\pi)^{-n/2} \int_{\mathbb{R}^n} e^{ix \cdot \xi} m(\xi) \hat{\varphi}(\xi) \, d\xi, \quad x \in \mathbb{R}^n.$$

Now, we can give a proof of Theorem 11.7.

Proof of Theorem 11.7 Let $\mathbb{Z}^n$ be the set of all n-tuples in $\mathbb{R}^n$ with integer coordinates. We write $\mathbb{R}^n$ as a union of cubes with disjoint interiors, i.e., $\mathbb{R}^n = \cup_{m \in \mathbb{Z}^n} Q_m$, where Q_m is the cube with center at m, edges of length one and parallel to the coordinate axes. Let Q_0 be the cube with center at the origin. Let η be any function in $C_0^\infty(\mathbb{R}^n)$ such that $\eta(x) = 1$ for all $x \in Q_0$. For $m \in \mathbb{Z}^n$, define σ_m by

$$\sigma_m(x, \xi) = \eta(x - m)\sigma(x, \xi), \quad x, \xi \in \mathbb{R}^n.$$

Obviously, $T_{\sigma_m} = \eta(x - m)T_\sigma$, and

$$\int_{Q_m} |(T_\sigma \varphi)(x)|^p dx \leq \int_{\mathbb{R}^n} |(T_{\sigma_m} \varphi)(x)|^p dx, \quad \varphi \in \mathcal{S}. \tag{11.5}$$

Since $\sigma_m(x, \xi)$ has compact support in x, it follows from Theorem 4.7 and Fubini's theorem that

$$\begin{aligned}
(T_{\sigma_m} \varphi)(x) &= (2\pi)^{-n/2} \int_{\mathbb{R}^n} e^{ix \cdot \xi} \sigma_m(x, \xi) \hat{\varphi}(\xi) \, d\xi \\
&= (2\pi)^{-n} \int_{\mathbb{R}^n} e^{ix \cdot \xi} \left\{ \int_{\mathbb{R}^n} e^{ix \cdot \lambda} \widehat{\sigma_m}(\lambda, \xi) \, d\lambda \right\} \hat{\varphi}(\xi) \, d\xi \\
&= (2\pi)^{-n} \int_{\mathbb{R}^n} e^{ix \cdot \lambda} \left\{ \int_{\mathbb{R}^n} e^{ix \cdot \xi} \widehat{\sigma_m}(\lambda, \xi) \hat{\varphi}(\xi) \, d\xi \right\} d\lambda, \tag{11.6}
\end{aligned}$$

where

$$\widehat{\sigma_m}(\lambda, \xi) = (2\pi)^{-n/2} \int_{\mathbb{R}^n} e^{-i\lambda \cdot x} \sigma_m(x, \xi) \, dx, \quad \lambda, \xi \in \mathbb{R}^n.$$

Lemma 11.9. *For all multi-indices α and positive integers N, there is a positive constant $C_{\alpha,N}$, depending on α and N only, such that*

$$|(D_\xi^\alpha \widehat{\sigma_m})(\lambda, \xi)| \leq C_{\alpha,N}(1 + |\xi|)^{-|\alpha|}(1 + |\lambda|)^{-N}, \quad \lambda, \xi \in \mathbb{R}^n.$$

The proof of Lemma 11.9, though easy, will be given later. This lemma and Theorem 11.8 imply that the operator $\varphi \mapsto T_\lambda \varphi$, defined on $\mathcal{S}$ by

$$(T_\lambda \varphi)(x) = (2\pi)^{-n/2} \int_{\mathbb{R}^n} e^{ix \cdot \xi} \widehat{\sigma_m}(\lambda, \xi) \hat{\varphi}(\xi) \, d\xi, \tag{11.7}$$

can be extended to a bounded linear operator on $L^p(\mathbb{R}^n)$. Moreover, for any positive integer N, there is a positive constant C_N such that

$$\|T_\lambda \varphi\|_p \leq C_N(1 + |\lambda|)^{-N} \|\varphi\|_p, \quad \varphi \in \mathcal{S}. \tag{11.8}$$

Using (11.6)–(11.8) and Minkowski's inequality in integral form,

$$\|T_{\sigma_m}\varphi\|_p = (2\pi)^{-n/2}\left\{\int_{\mathbb{R}^n}\left|\int_{\mathbb{R}^n}e^{ix\cdot\lambda}(T_\lambda\varphi)(x)\,d\lambda\right|^p dx\right\}^{1/p}$$

$$\leq (2\pi)^{-n/2}\int_{\mathbb{R}^n}\left\{\int_{\mathbb{R}^n}|(T_\lambda\varphi)(x)|^p dx\right\}^{1/p}d\lambda$$

$$= (2\pi)^{-n/2}\int_{\mathbb{R}^n}\|T_\lambda\varphi\|_p d\lambda$$

$$\leq C_N(2\pi)^{-n/2}\left\{\int_{\mathbb{R}^n}(1+|\lambda|)^{-N}d\lambda\right\}\|\varphi\|_p, \quad \varphi \in \mathcal{S}.$$

By choosing N sufficiently large, we can get another positive constant C_N such that

$$\|T_{\sigma_m}\varphi\|_p \leq C_N\|\varphi\|_p, \quad \varphi \in \mathcal{S}. \tag{11.9}$$

Hence, by (11.5) and (11.9),

$$\int_{Q_m}|(T_\sigma\varphi)(x)|^p dx \leq C_N^p\|\varphi\|_p^p, \quad \varphi \in \mathcal{S}. \tag{11.10}$$

Now, we represent T_σ as a singular integral operator. Precisely, we have

Lemma 11.10. *Let*

$$K(x,z) = (2\pi)^{-n/2}\int_{\mathbb{R}^n}e^{iz\cdot\xi}\sigma(x,\xi)d\xi$$

in the distribution sense. Then
(i) for each fixed $x \in \mathbb{R}^n$, $K(x,\cdot)$ is a function defined on $\mathbb{R}^n - \{0\}$,
(ii) for each sufficiently large positive integer N, there is a positive constant C_N such that

$$|K(x,z)| \leq C_N|z|^{-N}, \quad z \neq 0,$$

(iii) for each fixed $x \in \mathbb{R}^n$ and $\varphi \in \mathcal{S}$ vanishing on a neighborhood of x,

$$(T_\sigma\varphi)(x) = (2\pi)^{-n/2}\int_{\mathbb{R}^n}K(x,x-z)\varphi(z)\,dz.$$

Let us assume Lemma 11.10 for a moment. Let Q_m^{**} be the double of Q_m, i.e., Q_m^{**} has the same center as Q_m and edges parallel to the coordinate axes and twice the edge length of Q_m. Let Q_m^* be another cube concentric with Q_m and Q_m^{**} such that $Q_m \subset Q_m^* \subset Q_m^{**}$. Furthermore, we assume that there is a positive number δ such that $|x - z| \geq \delta$ for all $x \in Q_m$

and $z \in \mathbb{R}^n - Q_m^*$. The geometry is illustrated by the following figure.

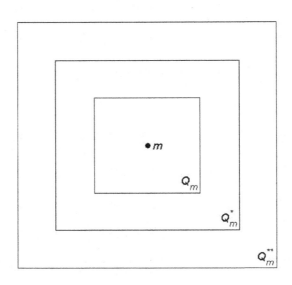

Let $\psi \in C_0^\infty(\mathbb{R}^n)$ be such that

$$0 \le \psi(x) \le 1, \quad x \in \mathbb{R}^n,$$

$$\text{supp}(\psi) \subseteq Q_m^{**},$$

and

$$\psi(x) = 1$$

on a neighborhood of Q_m^*. Write $\varphi = \varphi_1 + \varphi_2$, where $\varphi_1 = \psi\varphi$, and $\varphi_2 = (1 - \psi)\varphi$. Then

$$T_\sigma\varphi = T_\sigma\varphi_1 + T_\sigma\varphi_2.$$

Write

$$I_m = \int_{Q_m} |(T_\sigma\varphi)(x)|^p dx$$

and

$$J_m = \int_{Q_m} |(T_\sigma\varphi_2)(x)|^p dx.$$

Then for any sufficiently large positive integer N, inequality (11.10) implies that there is a positive constant C_N such that

$$
\begin{aligned}
I_m &= \int_{Q_m} |(T_\sigma \varphi_1)(x) + (T_\sigma \varphi_2)(x)|^p \, dx \\
&\leq 2^p \int_{Q_m} |(T_\sigma \varphi_1)(x)|^p \, dx + 2^p J_m \\
&\leq 2^p C_N^p \|\varphi_1\|_p^p + 2^p J_m.
\end{aligned}
\tag{11.11}
$$

By Lemma 11.10, there is a positive constant C_{2N} such that for all $x \in Q_m$,

$$
\begin{aligned}
|(T_\sigma \varphi_2)(x)| &= (2\pi)^{-n/2} \left| \int_{\mathbb{R}^n} K(x, x - z) \varphi_2(z) \, dz \right| \\
&= (2\pi)^{-n/2} \left| \int_{\mathbb{R}^n - Q_m^*} K(x, x - z) \varphi_2(z) \, dz \right| \\
&\leq C_{2N} \int_{\mathbb{R}^n - Q_m^*} |x - z|^{-2N} |\varphi_2(z)| \, dz.
\end{aligned}
\tag{11.12}
$$

Let $\lambda \geq \sqrt{n} + 1$. Then there exists a positive constant $C_{\lambda, N}$, depending on λ and N only, such that

$$
\frac{|x - z|^{-2N}}{(\lambda + |x - z|)^{-2N}} = \frac{(\lambda + |x - z|)^{2N}}{|x - z|^{2N}} \leq C_{\lambda, N}
\tag{11.13}
$$

for all $x \in Q_m$ and $z \in \mathbb{R}^n - Q_m^*$. Hence, by (11.12) and (11.13),

$$
\begin{aligned}
&|(T_\sigma \varphi_2)(x)| \\
&\leq C_{2N} C_{\lambda, N} \int_{\mathbb{R}^n - Q_m^*} (\lambda + |x - z|)^{-2N} |\varphi_2(z)| \, dz, \quad x \in Q_m.
\end{aligned}
\tag{11.14}
$$

Next, we note that for all $x \in Q_m$ and $z \in \mathbb{R}^n - Q_m^*$,

$$
\begin{aligned}
\lambda + |x - z| &= \lambda + |x - m + m - z| \\
&\geq \lambda + |m - z| - |x - m| \\
&\geq \left(\lambda - \frac{\sqrt{n}}{2} \right) + |m - z| \\
&\geq \mu + |m - z|,
\end{aligned}
\tag{11.15}
$$

where $\mu = \frac{\sqrt{n}}{2} + 1$. By (11.14) and (11.15),

$$
\begin{aligned}
&|(T_\sigma \varphi_2)(x)| \\
&\leq C_{2N} C_{\lambda, N} \int_{\mathbb{R}^n - Q_m^*} \frac{(\mu + |x - z|)^{-N} |\varphi_2(z)|}{(\mu + |m - z|)^N} \, dz, \quad x \in Q_m.
\end{aligned}
$$

By Minkowski's inequality in integral form and Hölder's inequality,

$$\left(\int_{Q_m} |(T_\sigma \varphi_2)(x)|^p dx\right)^{1/p}$$

$$\leq C_{2N} C_{\lambda N} \left\{\int_{Q_m} \left|\int_{\mathbb{R}^n - Q_m^*} \frac{(\mu + |x - z|)^{-N}|\varphi_2(z)|}{(\mu + |m - z|)^N} dz\right|^p dx\right\}^{1/p}$$

$$\leq C_{2N} C_{\lambda,N} \int_{\mathbb{R}^n - Q_m^*} \left\{\int_{Q_m} \frac{(\mu + |x - z|)^{-Np}|\varphi_2(z)|^p}{(\mu + |m - z|)^{Np}} dx\right\}^{1/p} dz$$

$$= C_{2N} C_{\lambda N} \int_{\mathbb{R}^n - Q_m^*} \frac{|\varphi_2(z)|}{(\mu + |m - z|)^N} \left\{\int_{Q_m} (\mu + |x - z|)^{-Np} dx\right\}^{1/p} dz$$

$$\leq C_{2N} C_{\lambda,N} \left\{\int_{\mathbb{R}^n - Q_m^*} (\mu + |m - z|)^{-Np'/2} dz\right\}^{1/p'}$$

$$\left\{\int_{\mathbb{R}^n - Q_m^*} \frac{|\varphi_2(z)|^p}{(\mu + |m - z|)^{Np/2}} dz\right\}^{1/p}.$$

Hence for any sufficiently large positive integer N, there is a positive constant $C_{\lambda,N,p}$, depending on λ, N and p only, such that

$$J_m \leq C_{\lambda,N,p} \int_{\mathbb{R}^n - Q_m^*} \frac{|\varphi_2(z)|^p}{(\mu + |m - z|)^{Np/2}} dz. \tag{11.16}$$

By (11.11) and (11.16),

$$I_m \leq 2^p C_N^p \int_{Q_m^{**}} |\varphi(x)|^p dx + 2^p C_{\lambda,N,p} \int_{\mathbb{R}^n - Q_m^*} \frac{|\varphi_2(z)|^p}{(\mu + |m - z|)^{Np/2}} dz.$$

Summing over all m in $\mathbb{Z}^n$, we get a positive constant C, depending only on n, p, N and λ, such that

$$\int_{\mathbb{R}^n} |(T_\sigma \varphi)(x)|^p dx$$

$$\leq 2^p C_N^p \sum_{m \in \mathbb{Z}^n} \int_{Q_m^{**}} |\varphi(x)|^p dx +$$

$$2^p C_{\lambda,N,p} \sum_{m \in \mathbb{Z}^n} \int_{\mathbb{R}^n - Q_m^*} \frac{|\varphi_2(z)|^p}{(\mu + |m - z|)^{Np/2}} dz$$

$$\leq C \int_{\mathbb{R}^n} |\varphi(x)|^p dx + 2^p C_{\lambda,N,p} \sum_{m \in \mathbb{Z}^n} \int_{\mathbb{R}^n - Q_m} \frac{|\varphi_2(z)|^p}{(\mu + |m - z|)^{Np/2}} dz$$

$$= C \int_{\mathbb{R}^n} |\varphi(x)|^p dx + 2^p C_{\lambda,N,p} \sum_{m \in \mathbb{Z}^n} \sum_{l \neq m} \int_{Q_l} \frac{|\varphi_2(z)|^p}{(\mu + |m - z|)^{Np/2}} dz.$$

$$\tag{11.17}$$

But using the same argument as in the derivation of (11.15), we get

$$\mu + |m - z| \geq 1 + |m - l| \tag{11.18}$$

for all $z \in Q_l$ and $l \neq m$. By (11.18),

$$\sum_{m \in \mathbb{Z}^n} \sum_{l \neq m} \int_{Q_l} \frac{|\varphi_2(z)|^p}{(\mu + |m - z|)^{Np/2}} dz$$

$$\leq \sum_{m \in \mathbb{Z}^n} \sum_{l \neq m} \frac{1}{(1 + |m - l|)^{Np/2}} \int_{Q_l} |\varphi_2(z)|^p dz$$

$$\leq \sum_{m \in \mathbb{Z}^n} \sum_{l \in \mathbb{Z}^n} \frac{1}{(1 + |m - l|)^{Np/2}} \int_{Q_l} |\varphi_2(z)|^p dz$$

$$= \sum_{l \in \mathbb{Z}^n} \int_{Q_l} |\varphi_2(z)|^p dz \sum_{m \in \mathbb{Z}^n} \frac{1}{(1 + |m - l|)^{Np/2}}$$

$$= \sum_{l \in \mathbb{Z}^n} \int_{Q_l} |\varphi_2(z)|^p dz \sum_{m \in \mathbb{Z}^n} \frac{1}{(1 + |m|)^{Np/2}}$$

$$= \sum_{m \in \mathbb{Z}^n} \frac{1}{(1 + |m|)^{Np/2}} \int_{\mathbb{R}^n} |\varphi_2(z)|^p dz. \tag{11.19}$$

Hence, by (11.17) and (11.19),

$$\int_{\mathbb{R}^n} |(T_\sigma \varphi)(x)|^p dx \leq \left\{ C + 2^p C_{\lambda, N, p} \sum_{m \in \mathbb{Z}^n} \frac{1}{(1 + |m|)^{Np/2}} \right\} \int_{\mathbb{R}^n} |\varphi(x)|^p dx.$$

Since $\mathcal{S}$ is dense in $L^p(\mathbb{R}^n)$ by Remark 3.10, it follows that T_σ can be extended to a bounded linear operator on $L^p(\mathbb{R}^n)$. $\qquad \square$

Remark 11.11. We leave it as an exercise to prove that the bounded extension coincides with $T_\sigma : \mathcal{S}' \to \mathcal{S}'$ restricted to the space $L^p(\mathbb{R}^n)$. See Exercise 11.2.

We now come to the proofs of Lemmas 11.9 and 11.10.

Proof of Lemma 11.9 Let β be an arbitrary multi-index. Then, by integration by parts and Leibniz's formula,

$$
(-i\lambda)^\beta (D_\xi^\alpha \widehat{\sigma_m})(\lambda, \xi)
$$

$$
= (-i\lambda)^\beta D_\xi^\alpha (2\pi)^{-n/2} \int_{\mathbb{R}^n} e^{-ix\cdot\lambda} \sigma_m(x, \xi)\, dx
$$

$$
= (-i\lambda)^\beta D_\xi^\alpha (2\pi)^{-n/2} \int_{\mathbb{R}^n} e^{-ix\cdot\lambda} \eta(x - m)\sigma(x, \xi)\, dx
$$

$$
= D_\xi^\alpha (2\pi)^{-n/2} \int_{\mathbb{R}^n} \{\partial_x^\beta e^{-ix\cdot\lambda}\} \eta(x - m)\sigma(x, \xi)\, dx
$$

$$
= (2\pi)^{-n/2} \int_{\mathbb{R}^n} \{\partial_x^\beta e^{-ix\cdot\lambda}\} \eta(x - m)(D_\xi^\alpha \sigma)(x, \xi)\, dx
$$

$$
= (-1)^{|\beta|} (2\pi)^{-n/2} \int_{\mathbb{R}^n} e^{-ix\cdot\lambda} \partial_x^\beta \{\eta(x - m)(D_\xi^\alpha \sigma)(x, \xi)\} dx
$$

$$
= (-1)^{|\beta|} (2\pi)^{-n/2} \sum_{\gamma \leq \beta} \binom{\beta}{\gamma} \int_{\mathbb{R}^n} e^{-ix\cdot\lambda} (\partial_x^\gamma \eta)(x - m)(\partial_x^{\beta-\gamma} D_\xi^\alpha \sigma)(x, \xi)\, dx.
$$

Using the properties of η and the fact that $\sigma \in S^0$, we can find a positive constant $C_{\alpha,\beta}$, depending on α and β only, such that

$$
|(-i\lambda)^\beta (D_\xi^\alpha \widehat{\sigma_m})(\lambda, \xi)| \leq C_{\alpha,\beta}(1 + |\xi|)^{-|\alpha|}, \quad \lambda, \xi \in \mathbb{R}^n.
$$

The lemma follows easily from this estimate. □

Proof of Lemma 11.10 Let α be an arbitrary multi-index with length greater than n. Then

$$
(-iz)^\alpha K(x, z) = (2\pi)^{-n/2} \int_{\mathbb{R}^n} e^{i\xi\cdot z} (\partial_\xi^\alpha \sigma)(x, \xi)\, d\xi \tag{11.20}
$$

in the distribution sense. Since $\sigma \in S^0$, it follows from (11.20) and Proposition 4.3 that $(iz)^\alpha K(x, z)$ is a continuous function on $\mathbb{R}^n$ and there is a positive constant C_α such that

$$
|z^\alpha|\, |K(x, z)| \leq C_\alpha
$$

for all $x, z \in \mathbb{R}^n$. Hence part (i) follows immediately and part (ii) follows if we use the inequality in Exercise 7.2. To prove part (iii), we define the tempered distribution L_x by

$$
L_x(\psi) = \int_{\mathbb{R}^n} \sigma(x, \xi)\psi(\xi)d\xi, \quad \psi \in \mathcal{S}.
$$

Then, by the definition of a pseudo-differential operator, Proposition 4.4 and the definition of the Fourier transform of a tempered distribution,

$$\begin{aligned}
(T_\sigma \varphi)(x) &= (2\pi)^{-n/2} \int_{\mathbb{R}^n} e^{ix\cdot\xi} \sigma(x,\xi) \hat{\varphi}(\xi)\, d\xi \\
&= (2\pi)^{-n/2} L_x(M_x \hat{\varphi}) \\
&= (2\pi)^{-n/2} L_x((T_x \varphi)^\wedge) \\
&= (2\pi)^{-n/2} \widehat{L_x}(T_x \varphi).
\end{aligned} \tag{11.21}$$

By part (i),

$$\widehat{L_x}(\psi) = \int_{\mathbb{R}^n} K(x,-z)\psi(z)\, dz \tag{11.22}$$

for all $\psi \in \mathcal{S}$ vanishing on a neighborhood of the origin. Hence, by (11.21) and (11.22),

$$\begin{aligned}
(T_\sigma \varphi)(x) &= (2\pi)^{-n/2} \int_{\mathbb{R}^n} K(x,-z)(T_x \varphi)(z)\, dz \\
&= (2\pi)^{-n/2} \int_{\mathbb{R}^n} K(x,-z)\varphi(x+z)\, dz \\
&= (2\pi)^{-n/2} \int_{\mathbb{R}^n} K(x,x-z)\varphi(z)\, dz
\end{aligned}$$

and the proof is complete. $\qquad\square$

Remark 11.12. The proof of the L^p-boundedness of pseudo-differential operators for $1 < p < \infty$ given in Theorem 11.7 is based on the Hörmander multiplier theorem, which is formulated without proof as Theorem 11.8. All proofs of the alluded L^p-boundedness invoke some results and concepts that go beyond the elementary nature of this book. For L^2-boundedness, we can provide a complete and self-contained proof. Indeed, for all positive integers N, we obtain by Lemma 11.9 a positive constant C_N such that

$$|\widehat{\sigma_m}(\lambda,\xi)| \le C_N(1+|\lambda|)^{-N}, \quad \lambda, \xi \in \mathbb{R}^n.$$

By (11.7) and the Plancherel theorem, we get for all positive integers N, a positice constant C_N such that

$$\|T_\lambda \varphi\|_2 = \|\widehat{\sigma_m}(\lambda,\cdot)\hat{\varphi}\|_2 \le C_N(1+|\lambda|)^{-N}\|\hat{\varphi}\|_2 = C_N(1+|\lambda|)^{-N}\|\varphi\|_2,$$

which is (11.8) for $p = 2$. The use of Hörmander's multiplier theorem is thus avoided.

Exercises

11.1. Prove that the definition of $T_\sigma u$ given for a tempered distribution u given in (11.4) coincides with the definition of $T_\sigma u$ for a Schwartz function u given in (6.4).

11.2. Let $\sigma \in S^0$ and $1 < p < \infty$. Then we have shown in the proof of Theorem 10.7 that there is a positive constant C such that

$$\|T_\sigma \varphi\|_p \le C \|\varphi\|_p, \quad \varphi \in \mathcal{S}.$$

Hence T_σ can be extended to a unique bounded linear operator from $L^p(\mathbb{R}^n)$ into $L^p(\mathbb{R}^n)$. Prove that the extension coincides with $T_\sigma : \mathcal{S}' \to \mathcal{S}'$ restricted to $L^p(\mathbb{R}^n)$.

11.3. Let $\sigma(x,\xi) \in S^0$ be a nonzero symbol which is independent of $\xi \in \mathbb{R}^n$. Prove that the bounded linear operator $T_\sigma : L^p(\mathbb{R}^n) \to L^p(\mathbb{R}^n)$, $1 < p < \infty$, is not compact.

11.4. Show that the limit in operator norm of a sequence of pseudo-differential operators $T_{\sigma_k} : L^2(\mathbb{R}^n) \to L^2(\mathbb{R}^n)$, where $\sigma_k \in S^0$, need not be a pseudo-differential operator.

11.5. Suppose that $\sigma \in S^0$ has compact support in x. Prove that the pseudo-differential operator $T_\sigma : L^p(\mathbb{R}^n) \to L^p(\mathbb{R}^n)$ is a bounded linear operator for $1 < p < \infty$.

Chapter 12

The Sobolev Spaces
$H^{s,p}$, $-\infty < s < \infty$, $1 \leq p < \infty$

Theorem 11.7 tells us that $T_\sigma : L^p(\mathbb{R}^n) \to L^p(\mathbb{R}^n)$ is a bounded linear operator for $1 < p < \infty$ if σ is a symbol in S^0. In order to find an analog of Theorem 11.7 for an arbitrary symbol in S^m, we need to introduce a family of spaces of tempered distributions.

For $-\infty < s < \infty$, we denote by J_s the pseudo-differential operator of which the symbol $\sigma_s(\xi)$ is given by

$$\sigma_s(\xi) = (1 + |\xi|^2)^{-s/2}, \quad \xi \in \mathbb{R}^n.$$

It should be noted that the symbol of J_s is in S^{-s}. (See Example 6.4.) The operator J_s is often called the *Bessel potential of order s*.

It is an easy exercise to prove that for any $u \in \mathcal{S}'$, the product $\sigma_s u$ of σ_s and u defined by

$$(\sigma_s u)(\varphi) = u(\sigma_s \varphi), \quad \varphi \in \mathcal{S},$$

is also in $\mathcal{S}'$. See Exercise 12.1.

The following proposition is an easy consequence of (11.3) and the definition of J_s, and the proof is left as an exercise. See Exercise 12.2.

Proposition 12.1. $J_s u = \mathcal{F}^{-1} \sigma_s \mathcal{F} u, \quad u \in \mathcal{S}'.$

An easy corollary of Proposition 12.1 is the following proposition. Its proof is also left as an exercise. See Exercise 12.3.

Proposition 12.2. *Let $u \in \mathcal{S}'$. Then*
(i) $J_s J_t u = J_{s+t} u$,
(ii) $J_0 u = u$.

For $-\infty < s < \infty$ and $1 \leq p < \infty$, we define $H^{s,p}$ to be the set of all tempered distributions u for which $J_{-s} u$ is a function in $L^p(\mathbb{R}^n)$. It is obvious that $H^{s,p}$ is a vector space. It can be made into a normed vector space if we equip it with the norm $\| \|_{s,p}$, where

$$\|u\|_{s,p} = \|J_{-s}u\|_p, \quad u \in H^{s,p}.$$

We usually call $H^{s,p}$ the *L^p-Sobolev space of order s*. It is obvious that $H^{0,p} = L^p(\mathbb{R}^n)$.

Theorem 12.3. *$H^{s,p}$ is a Banach space with respect to the norm $\| \|_{s,p}$.*

Proof We need only prove completeness. To do this, let $\{u_k\}$ be a Cauchy sequence in $H^{s,p}$. Then, by the definition of $H^{s,p}$, the sequence $\{J_{-s}u_k\}$ is a Cauchy sequence in $L^p(\mathbb{R}^n)$. Since $L^p(\mathbb{R}^n)$ is complete, it follows that there exists a function u in $L^p(\mathbb{R}^n)$ such that

$$J_{-s}u_k \to u \tag{12.1}$$

in $L^p(\mathbb{R}^n)$ as $k \to \infty$. Let $v = J_s u$. Then, by Proposition 12.2, $J_{-s}v = u$. Hence $J_{-s}v$ is in $L^p(\mathbb{R}^n)$, i.e., $v \in H^{s,p}$. That $u_k \to v$ in $H^{s,p}$ as $k \to \infty$ is an immediate consequence of (12.1). $\quad\square$

Proposition 12.4. *J_t is an isometry of $H^{s,p}$ onto $H^{s+t,p}$. More precisely,*

$$\|J_t u\|_{s+t,p} = \|u\|_{s,p}, \quad u \in H^{s,p}. \tag{12.2}$$

Proof Let $u \in H^{s,p}$. Then, by Proposition 12.2,

$$\|J_t u\|_{s+t,p} = \|J_{-s-t}J_t u\|_p = \|J_{-s}u\|_p = \|u\|_{s,p}.$$

Let $v \in H^{s+t,p}$. Then, by Proposition 12.2, $J_{-t}v \in H^{s,p}$ and $J_t J_{-t}v = v$. This proves that J_t is onto. $\quad\square$

Theorem 12.5. *Let $1 \le p < \infty$ and $s \le t$. Then $H^{t,p} \subseteq H^{s,p}$, and*

$$\|u\|_{s,p} \le \|u\|_{t,p}, \quad u \in H^{t,p}.$$

Theorem 12.5 is usually known as the *Sobolev embedding theorem*. To prove Theorem 12.5, we use a technical result which gives us an explicit formula for the inverse Fourier transform of the function $(1 + |\xi|^2)^{-s/2}$, where $\xi \in \mathbb{R}^n$ and $s > 0$, in the distribution sense.

Proposition 12.6. *Let $s > 0$. If we define the function G_s on $\mathbb{R}^n$ by*

$$G_s(x) = \frac{1}{2^{s/2}\Gamma\left(\frac{s}{2}\right)} \int_0^\infty e^{-r/2}e^{-|x|^2/(2r)}r^{-(n-s)/2}\frac{dr}{r}, \quad x \in \mathbb{R}^n,$$

then

(i) $G_s \in L^1(\mathbb{R}^n)$,

(ii) $\|G_s\|_1 = (2\pi)^{n/2}$,

(iii) $\widehat{G_s}(\xi) = \left(1 + |\xi|^2\right)^{-s/2}, \quad \xi \in \mathbb{R}^n.$

Proof By Fubini's theorem,

$$
\int_{\mathbb{R}^n} |G_s(x)|\, dx = \int_{\mathbb{R}^n} G_s(x)\, dx
$$

$$
= \frac{1}{2^{s/2}\Gamma\left(\frac{s}{2}\right)} \int_{\mathbb{R}^n} \left\{ \int_0^\infty e^{-r/2} e^{-|x|^2/(2r)} r^{-(n-s)/2} \frac{dr}{r} \right\} dx
$$

$$
= \frac{1}{2^{s/2}\Gamma\left(\frac{s}{2}\right)} \int_0^\infty e^{-r/2} r^{-(n-s)/2} \left\{ \int_{\mathbb{R}^n} e^{-|x|^2/(2r)} dx \right\} \frac{dr}{r}
$$

$$
= \frac{(2\pi)^{n/2}}{2^{s/2}\Gamma\left(\frac{s}{2}\right)} \int_0^\infty e^{-r/2} r^{s/2} \frac{dr}{r} \tag{12.3}
$$

if we recall that

$$
\int_{\mathbb{R}^n} e^{-|x|^2/(2r)} dx = (2\pi r)^{n/2}.
$$

But for any $\varepsilon > 0$ and $a > 0$, we have

$$
\varepsilon^{-a}\Gamma(a) = \int_0^\infty e^{-\varepsilon r} r^a \frac{dr}{r}. \tag{12.4}
$$

Putting $\varepsilon = \frac{1}{2}$ and $a = \frac{s}{2}$ in (12.4), we get

$$
\int_0^\infty e^{-r/2} r^{s/2} \frac{dr}{r} = 2^{s/2}\Gamma\left(\frac{s}{2}\right). \tag{12.5}
$$

Hence, by (12.3) and (12.5),

$$
\int_{\mathbb{R}^n} |G_s(x)| dx = (2\pi)^{n/2}.
$$

This proves parts (i) and (ii). To prove part (iii), let $\varphi \in \mathcal{S}$. Then, by Proposition 4.6,

$$
\int_{\mathbb{R}^n} \widehat{G}_s(\xi)\varphi(\xi)\, d\xi = \int_{\mathbb{R}^n} G_s(\xi)\hat{\varphi}(\xi)\, d\xi
$$

$$
= \frac{1}{2^{s/2}\Gamma\left(\frac{s}{2}\right)} \int_{\mathbb{R}^n} \left\{ \int_0^\infty e^{-r/2} e^{-|\xi|^2/(2r)} r^{-(n-s)/2} \frac{dr}{r} \right\} \hat{\varphi}(\xi)\, d\xi.
$$

Using Fubini's theorem, we get

$$
\int_{\mathbb{R}^n} \widehat{G}_s(\xi)\varphi(\xi)\, d\xi
$$

$$
= \frac{1}{2^{s/2}\Gamma\left(\frac{s}{2}\right)} \int_0^\infty e^{-r/2} r^{-(n-s)/2} \left\{ \int_{\mathbb{R}^n} \hat{\varphi}(\xi) e^{-|\xi|^2/(2r)} d\xi \right\} \frac{dr}{r}. \tag{12.6}
$$

But, by Proposition 4.6 again, we get

$$
\int_{\mathbb{R}^n} \hat{\varphi}(\xi) e^{-|\xi|^2/(2r)} d\xi = \int_{\mathbb{R}^n} \varphi(\xi)\hat{\psi}(\xi)\, d\xi, \tag{12.7}
$$

where

$$\psi(x) = e^{-|x|^2/(2r)}, \quad x \in \mathbb{R}^n.$$

By Propositions 4.4 (iii) and 4.5,

$$\hat{\psi}(\xi) = r^{n/2}e^{-r|\xi|^2/2}, \quad \xi \in \mathbb{R}^n.$$

Hence, by (12.7),

$$\int_{\mathbb{R}^n} \hat{\varphi}(\xi)e^{-|\xi|^2/(2r)}d\xi = r^{n/2} \int_{\mathbb{R}^n} \varphi(\xi)e^{-r|\xi|^2/2}d\xi. \tag{12.8}$$

Therefore, by (12.6) and (12.8),

$$\int_{\mathbb{R}^n} \widehat{G_s}(\xi)\varphi(\xi)\, d\xi = \frac{1}{2^{s/2}\Gamma\left(\frac{s}{2}\right)} \int_0^\infty e^{-r/2}r^{s/2} \left\{ \int_{\mathbb{R}^n} \varphi(\xi)e^{-r|\xi|^2/2}d\xi \right\} \frac{dr}{r}.$$

Using Fubini's theorem again, we get

$$\int_{\mathbb{R}^n} \widehat{G_s}(\xi)\varphi(\xi)\, d\xi$$

$$= \frac{1}{2^{s/2}\Gamma\left(\frac{s}{2}\right)} \int_{\mathbb{R}^n} \varphi(\xi) \left\{ \int_0^\infty e^{-r/2}r^{s/2}e^{-r|\xi|^2/2}\frac{dr}{r} \right\} d\xi$$

$$= \frac{1}{2^{s/2}\Gamma\left(\frac{s}{2}\right)} \int_{\mathbb{R}^n} \varphi(\xi) \left\{ \int_0^\infty e^{-r(1+|\xi|^2)/2}r^{s/2}\frac{dr}{r} \right\} d\xi.$$

Putting $\varepsilon = (1 + |\xi|^2)/2$ and $a = \frac{s}{2}$ in (12.4), we get

$$\int_{\mathbb{R}^n} \widehat{G_s}(\xi)\varphi(\xi)\, d\xi$$

$$= \frac{1}{2^{s/2}\Gamma\left(\frac{s}{2}\right)}\Gamma\left(\frac{s}{2}\right)2^{s/2} \int_{\mathbb{R}^n} \varphi(\xi)(1 + |\xi|^2)^{-s/2}d\xi$$

$$= \int_{\mathbb{R}^n} \left(1 + |\xi|^2\right)^{-s/2} \varphi(\xi)\, d\xi. \tag{12.9}$$

Hence, by Lemma 6.6 and (12.9), we can conclude that

$$\widehat{G_s}(\xi) = (1 + |\xi|^2)^{-s/2}, \quad \xi \in \mathbb{R}^n.$$

$\square$

The following consequence of Proposition 12.6 will be useful to us.

Proposition 12.7. *Let $s \geq 0$ and $1 \leq p < \infty$. Then*

$$\|J_s u\|_p \leq \|u\|_p, \quad u \in L^p(\mathbb{R}^n). \tag{12.10}$$

Proof Let $\varphi \in \mathcal{S}$. Then, by the definition of J_s,

$$(J_s\varphi)^\wedge(\xi) = (1 + |\xi|^2)^{-s/2}\hat\varphi(\xi), \quad \xi \in \mathbb{R}^n.$$

On the other hand, by Propositions 4.1 and 12.6,

$$(G_s * \varphi)^\wedge(\xi) = (2\pi)^{n/2}\widehat{G_s}(\xi)\hat\varphi(\xi)$$
$$= (2\pi)^{n/2}(1 + |\xi|^2)^{-s/2}\hat\varphi(\xi), \quad \xi \in \mathbb{R}^n.$$

Hence for all $\varphi \in \mathcal{S}$,

$$J_s\varphi = (2\pi)^{-n/2}(G_s * \varphi),$$

and, by Theorem 3.1,

$$\|J_s\varphi\|_p \le (2\pi)^{-n/2}\|G_s\|_1 \|\varphi\|_p = \|\varphi\|_p.$$

Since $\mathcal{S}$ is dense in $L^p(\mathbb{R}^n)$ by Remark 3.10, it follows that J_s can be extended to a bounded linear operator on $L^p(\mathbb{R}^n)$ satisfying (12.10).   □

Remark 12.8. As in Remark 11.11, we leave it as an exercise to prove that the bounded extension coincides with $J_s : \mathcal{S}' \to \mathcal{S}'$ restricted to the space $L^p(\mathbb{R}^n)$.

Proof of Theorem 12.5 Let $u \in H^{t,p}$. Then, by the definition of $H^{t,p}$, we have $J_{-t}u \in L^p(\mathbb{R}^n)$. Hence, by Proposition 12.2, $J_{-s}u = J_{t-s}J_{-t}u$. By the definition of $H^{s,p}$ and Proposition 12.7, we get

$$\|u\|_{s,p} = \|J_{-s}u\|_p = \|J_{t-s}J_{-t}u\|_p \le \|J_{-t}u\|_p = \|u\|_{t,p},$$

and hence Theorem 12.5 follows.   □

We can now give a more precise result generalizing Theorem 11.7.

Theorem 12.9. *Let σ be a symbol in S^m. Then $T_\sigma : H^{s,p} \to H^{s-m,p}$ is a bounded linear operator for $-\infty < s < \infty$ and $1 < p < \infty$.*

Proof Since $J_{m-s}T_\sigma J_s \in S^0$, it follows from Theorem 11.7 that there is a positive constant C such that

$$\|T_\sigma u\|_{s-m,p} = \|J_{m-s}T_\sigma u\|_p = \|J_{m-s}T_\sigma J_s J_{-s}u\|_p$$
$$\le C\|J_{-s}u\|_p = C\|u\|_{s,p}, \quad u \in H^{s,p}.$$

□

Remark 12.10. A weaker result than the Sobolev embedding theorem formulated in Theorem 12.5 is one that asserts the existence of a positive constant C such that

$$\|u\|_{s,p} \le C\|u\|_{t,p}, \quad u \in H^{t,p},$$

where $1 < p < \infty$. The weaker inequality has a much shorter proof. Indeed, for all $u \in H^{t,p}$,

$$\|u\|_{s,p} = \|J_{-s}u\|_p = \|J_{t-s}J_{-t}u\|_p.$$

Since $s \leq t$, J_{t-s} is a pseudo-differential operator with symbol in S^0. So, by Theorem 11.7, there exists a positive constant C such that

$$\|J_{-s}u\|_p = \|J_{t-s}J_{-t}u\|_p \leq C\|J_{-t}u\|_p = C\|u\|_{t,p}, \quad u \in H^{t,p}.$$

Thus,

$$\|u\|_{s,p} \leq \|u\|_{t,p}, \quad u \in H^{t,p}.$$

Notwithstanding the apparently shorter proof, the L^p-boundedness of pseudo-differential operators is invoked. So, the proof of Theorem 12.5 as given in this chapter is self-contained and elementary, and it gives a better result.

Exercises

12.1. Let $s \in (-\infty, \infty)$ and σ_s be the function on $\mathbb{R}^n$ defined by

$$\sigma_s(\xi) = (1 + |\xi|^2)^{-s/2}, \quad \xi \in \mathbb{R}^n.$$

Prove that if $u \in \mathcal{S}'$, then the product $\sigma_s u$ of σ_s and u defined by

$$(\sigma_s u)(\varphi) = u(\sigma_s \varphi), \quad \varphi \in \mathcal{S},$$

is also in $\mathcal{S}'$.

12.2. Prove that for all $s \in (-\infty, \infty)$,

$$J_s u = \mathcal{F}^{-1}\sigma_s \mathcal{F}u, \quad u \in \mathcal{S}',$$

where σ_s is the function on $\mathbb{R}^n$ defined in Exercise 12.1.

12.3. (i) Prove that for all $s, t \in (-\infty, \infty)$,

$$J_s J_t u = J_{s+t}u, \quad u \in \mathcal{S}'.$$

(ii) Prove that $J_0 u = u$ for all $u \in \mathcal{S}'$.

12.4. (i) Prove that for all $s \in (-\infty, \infty)$,

$$H^{s,2} = \left\{ u \in \mathcal{S}' : \int_{\mathbb{R}^n} (1 + |\xi|^2)^s |\hat{u}(\xi)|^2 d\xi < \infty \right\}.$$

(ii) Prove that for all $s \in (-\infty, \infty)$,

$$\|u\|_{s,2} = \left\{ \int_{\mathbb{R}^n} (1 + |\xi|^2)^s |\hat{u}(\xi)|^2 d\xi \right\}^{1/2}, \quad u \in H^{s,2}.$$

12.5. (i) Prove that if $u \in H^{s,2}$, $s > \frac{n}{2} + k$, where k is a nonnegative integer, then u is equal to a C^k function on $\mathbb{R}^n$ almost everywhere.

(ii) Prove that if $u \in H^{s,2}$, $s > \frac{n}{2}$, then u can be modified on a set of measure zero to a continuous function v on $\mathbb{R}^n$ such that

$$\lim_{|x| \to \infty} v(x) = 0.$$

12.6. (Erhling's Inequality) Prove that if $s < t$, then for any positive number ε, there exists a positive constant C, depending on ε, s and t only, such that

$$\|\varphi\|_{s,2} < \varepsilon \|\varphi\|_{t,2} + C \|\varphi\|_{0,2}, \quad \varphi \in \mathcal{S}.$$

12.7. Let $s \ge 0$ and $1 \le p < \infty$. Then we have shown in the proof of Proposition 12.7 that

$$\|J_s \varphi\|_p \le \|\varphi\|_p, \quad \varphi \in \mathcal{S}.$$

Hence J_s can be extended to a unique bounded linear operator from $L^p(\mathbb{R}^n)$ into $L^p(\mathbb{R}^n)$. Prove that the extension coincides with $J_s : \mathcal{S}' \to \mathcal{S}'$ restricted to $L^p(\mathbb{R}^n)$.

12.8. Let σ be an elliptic symbol in S^m. Let $u \in L^p(\mathbb{R}^n)$ be a solution of the pseudo-differential equation $T_\sigma u = f$, where $f \in L^p(\mathbb{R}^n)$. Prove that $u \in H^{m,p}$.

12.9. Let $\sigma \in S^m$, $m > 0$, and $f \in L^p(\mathbb{R}^n)$, $1 < p < \infty$. An *approximate solution* of the pseudo-differential equation $T_\sigma u = f$ on $\mathbb{R}^n$ is a function u in $H^{m,p}$ such that $T_\sigma u = f$ modulo $\cap_{s \in \mathbb{R}} H^{s,p}$, i.e., $T_\sigma u - f \in \cap_{s \in \mathbb{R}} H^{s,p}$. Prove that an approximate solution exists if σ is elliptic.

12.10. Find all real numbers s such that $\delta \in H^{s,2}$.

12.11. Let $s \in (-\infty, \infty)$. Find $J_s \delta$.

12.12. Let $s > 0$. Find a solution u in $L^1(\mathbb{R}^n)$ such that $(I - \Delta)^{s/2} u = \delta$ on $\mathbb{R}^n$, where $(I - \Delta)^{s/2}$ is the operator introduced in Example 6.4. (u is termed a *fundamental solution* of the operator $(I - \Delta)^{s/2}$.)

12.13. Is $\cap_{s \in \mathbb{R}} H^{s,2} = \mathcal{S}$? Explain your answer.

12.14. For $-\infty < s < \infty$ and $1 < p < \infty$, let $u \in H^{s,p}$ and $v \in H^{-s,p'}$. Let $\{\varphi_j\}$ and $\{\psi_j\}$ be sequences in $\mathcal{S}$ such that $\varphi_j \to u$ in $H^{s,p}$ and $\psi_j \to v$ in $H^{-s,p'}$ as $j \to \infty$.

(i) Prove that $\lim_{j\to\infty}(\varphi_j, \psi_j)$ exists and the limit is independent of the choice of the sequences $\{\varphi_j\}$ and $\{\psi_j\}$. (The limit is denoted by (u, v).)

(ii) Prove that

$$|(u,v)| \leq \|u\|_{s,p}\|v\|_{-s,p'}, \quad u \in H^{s,p}, \, v \in H^{-s,p'}.$$

Chapter 13

Closed Linear Operators

In this chapter we give a brief account of the theory of closed linear operators on Banach spaces. The choice of topics is dictated by what we need for the theory of minimal and maximal pseudo-differential operators in the next chapter.

Let X and Y be complex Banach spaces with norms denoted by $\| \; \|_X$ and $\| \; \|_Y$ respectively. We are interested in linear operators A mapping a dense subspace of X, usually denoted by $\mathcal{D}(A)$, into Y. We call $\mathcal{D}(A)$ the *domain* of the operator A.

Definition 13.1. The operator A is said to be *closed* if for any sequence $\{x_k\}$ of vectors in $\mathcal{D}(A)$ such that $x_k \to x$ in X and $Ax_k \to y$ in Y as $k \to \infty$, we have $x \in \mathcal{D}(A)$ and $Ax = y$.

Definition 13.2. The operator A is said to be *closable* if for any sequence $\{x_k\}$ of vectors in $\mathcal{D}(A)$ such that $x_k \to 0$ in X and $Ax_k \to y$ in Y as $k \to \infty$, we have $y = 0$.

Obviously, a closed linear operator is closable.

Definition 13.3. Let A and B be linear operators from X into Y with domains $\mathcal{D}(A)$ and $\mathcal{D}(B)$ respectively. We call B an *extension* of A if $\mathcal{D}(A) \subseteq \mathcal{D}(B)$ and $Bx = Ax$ for all $x \in \mathcal{D}(A)$.

Proposition 13.4. *Let A be any linear operator from X into Y with domain $\mathcal{D}(A)$. Then A has a closed extension if and only if A is closable.*

Proof Let B be a closed extension of A. Let $\{x_k\}$ be a sequence of vectors in $\mathcal{D}(A)$ such that $x_k \to 0$ in X and $Ax_k \to y$ in Y as $k \to \infty$. Since B is an extension of A, it follows that $x_k \in \mathcal{D}(B), x_k \to 0$ in X and $Bx_k \to y$ in Y as $k \to \infty$. Since B is closed, we have $y = 0$. Hence A is closable.

Conversely, suppose that A is closable. We define an operator A_0 as follows. $\mathcal{D}(A_0)$ is the set of all vectors $x \in X$ such that there is a sequence $\{x_k\}$ of vectors in $\mathcal{D}(A)$ with the property that $x_k \to x$ in X and $Ax_k \to y$ in Y for some $y \in Y$ as $k \to \infty$. For any $x \in \mathcal{D}(A_0)$, we define $A_0 x$ to be equal to y. We have to check that the definition of A_0 does not depend on the particular choice of the sequence $\{x_k\}$. Indeed, if $\{z_k\}$ is another sequence of vectors in $\mathcal{D}(A)$ such that $z_k \to x$ in X and $Az_k \to w$ for some other $w \in Y$ as $k \to \infty$, then $x_k - z_k \to 0$ in X as $k \to \infty$, $x_k - z_k \in \mathcal{D}(A)$, and $A(x_k - z_k) \to y - w$ in Y as $k \to \infty$. Since A is closable, it follows that $y - w = 0$, i.e., $y = w$. Obviously, A_0 is an extension of A. It is closed. For let $\{x_k\}$ be a sequence of vectors in $\mathcal{D}(A_0)$ such that $x_k \to x$ in X and $A_0 x_k \to y$ in Y as $k \to \infty$. For each k, there is a sequence $\{x_{k_j}\}$ of vectors in $\mathcal{D}(A)$ such that $x_{k_j} \to x_k$ in X and $Ax_{k_j} \to A_0 x_k$ in Y as $j \to \infty$. Hence for each k, there is a $z_k \in \mathcal{D}(A)$ such that

$$\|x_k - z_k\|_X < \frac{1}{k}$$

and

$$\|A_0 x_k - Az_k\|_Y < \frac{1}{k}.$$

Hence for each k,

$$\|z_k - x\|_X \leq \|x_k - z_k\|_X + \|x_k - x\|_X < \frac{1}{k} + \|x_k - x\|_X.$$

Therefore $\|z_k - x\|_X \to 0$ as $k \to \infty$. Similarly, $\|Az_k - y\|_Y \to 0$ as $k \to \infty$. This proves that $x \in \mathcal{D}(A_0)$ and $A_0 x = y$. Hence A_0 is closed. $\square$

Let us study the operator A_0 constructed in the proof of Proposition 13.4.

Proposition 13.5. *A_0 is the smallest closed extension of A. This means that if B is any closed extension of A, then B is an extension of A_0.*

Proof Let $x \in \mathcal{D}(A_0)$ and $A_0 x = y$. Then, by the definition of A_0, we can find a sequence $\{x_k\}$ of vectors in $\mathcal{D}(A)$ such that $x_k \to x$ in X and $Ax_k \to y$ in Y as $k \to \infty$. Since B is an extension of A, it follows that $x_k \in \mathcal{D}(B)$, $x_k \to x$ in X and $Bx_k \to y$ in Y as $k \to \infty$. Since B is closed, we can conclude that $x \in \mathcal{D}(B)$ and $Bx = y$. This proves that B is an extension of A_0. $\square$

Remark 13.6. In view of Proposition 13.5, we call A_0 the *minimal operator* of A.

Let X be any complex Banach space with norm $\| \ \|_X$. We denote by X' the dual space of X. Let us recall that X' is the Banach space of all bounded conjugate linear functionals on X. The norm $\| \ \|_{X'}$ in X' is given by

$$\|f\|_{X'} = \sup_{\substack{x \in X \\ x \neq 0}} \frac{|f(x)|}{\|x\|_X}, \quad f \in X'.$$

Let X and Y be complex Banach spaces. For any linear operator A from X into Y with domain $\mathcal{D}(A)$ *dense* in X, we define an operator $A^t : Y' \to X'$ as follows.

$\mathcal{D}(A^t)$ is the set of all functionals y' in Y' for which there is a functional x' in X' such that

$$y'(Ax) = x'(x), \quad x \in \mathcal{D}(A). \tag{13.1}$$

Lemma 13.7. *Let $y' \in Y'$. Then there is at most one $x' \in X'$ for which* (13.1) *holds.*

By Lemma 13.7, we can define $A^t y'$ to be equal to x' for all $y' \in \mathcal{D}(A^t)$. We call A^t the true adjoint or simply the *adjoint* of A.

Proof of Lemma 13.7 Let $y' \in Y'$. Suppose x' and z' are functionals in X' for which

$$y'(Ax) = x'(x), \quad x \in \mathcal{D}(A),$$

and

$$y'(Ax) = z'(x), \quad x \in \mathcal{D}(A).$$

Obviously, $x' = z'$ on $\mathcal{D}(A)$. Since $\mathcal{D}(A)$ is dense in X, a simple limiting argument will show that $x' = z'$ on X. $\qquad \square$

Proposition 13.8. *A^t is a closed linear operator from Y' into X'.*

Proof Linearity, as usual, is easy to check. To prove that A^t is closed, let $\{y_k'\}$ be a sequence of functionals in $\mathcal{D}(A^t)$ such that $y_k' \to y'$ in Y' and $A^t y_k' \to x'$ in X' as $k \to \infty$. Then, by (13.1) and the definition of A^t,

$$y_k'(Ax) = (A^t y_k')(x)$$

for all $x \in \mathcal{D}(A)$ and $k = 1, 2, \ldots$. Let $k \to \infty$. Then

$$y'(Ax) = x'(x), \quad x \in \mathcal{D}(A).$$

This proves that $y' \in \mathcal{D}(A^t)$ and $A^t y' = x'$. Therefore A^t is a closed operator. □

Another observation about the adjoint of a linear operator we shall use is given in the following proposition.

Proposition 13.9. *Let A be any linear operator from X into Y with domain $\mathcal{D}(A)$ dense in X. Then for any extension B of A, the operator A^t is an extension of B^t.*

Proof Let $y' \in \mathcal{D}(B^t)$. Then we can find a functional x' in X' such that

$$y'(Bx) = x'(x), \quad x \in \mathcal{D}(B).$$

Since B is an extension of A, we have

$$y'(Ax) = x'(x), \quad x \in \mathcal{D}(A).$$

This proves that $y' \in \mathcal{D}(A^t)$ and $A^t y' = x' = B^t y'$. □

Exercises

13.1. Prove that for $-\infty < s < \infty$ and $1 < p < \infty$, the dual space of $H^{s,p}$ is $H^{-s,p'}$, where

$$\frac{1}{p} + \frac{1}{p'} = 1.$$

13.2. Let A be a linear operator from $L^p(\mathbb{R}^n)$ into $L^p(\mathbb{R}^n)$, $1 < p < \infty$, with dense domain $\mathcal{D}(A)$. Prove that $\mathcal{D}(A^t)$ consists of all functions u in $L^{p'}(\mathbb{R}^n)$ for which there exists a function f in $L^{p'}(\mathbb{R}^n)$ such that

$$(u, Av) = (f, v), \quad v \in \mathcal{D}(A),$$

where

$$(g, h) = \int_{\mathbb{R}^n} g(x)\overline{h(x)}\, dx$$

for all $g \in L^{p'}(\mathbb{R}^n)$ and $h \in L^p(\mathbb{R}^n)$.

13.3. Let Z be a complex Banach space. A subset S of Z' is said to be *total* if

$$\{z \in Z : z'(z) = 0 \text{ for all } z' \in S\} = \{0\}.$$

Let A be a linear operator from a complex Banach space into another complex Banach space with dense domain $\mathcal{D}(A)$. Prove that A is closable if and only if the domain $\mathcal{D}(A^t)$ of A^t is a total set.

Chapter 14

Minimal and Maximal
Pseudo-Differential Operators

Let σ be a symbol in S^m. Then the pseudo-differential operator T_σ, initially defined in Chapter 6 on the Schwartz space $\mathcal{S}$, has later been extended in Chapter 11 to the space $\mathcal{S}'$ of all tempered distributions using the formal adjoint T_σ^*. By Theorem 12.9, $T_\sigma : H^{s,p} \to H^{s-m,p}$ is a bounded linear operator for $-\infty < s < \infty$ and $1 < p < \infty$. As a matter of fact, when $m > 0$, the operator T_σ can also be considered as a linear operator from $L^p(\mathbb{R}^n)$ into $L^p(\mathbb{R}^n)$, $1 < p < \infty$, with domain $\mathcal{S}$. We denote this operator simply by T_σ. It is not closed in general. Fortunately, it is closable. Hence, by Proposition 13.4, it has a closed extension.

Proposition 14.1. *The operator T_σ is closable.*

Proof Let $\{\varphi_k\}$ be a sequence of functions in $\mathcal{S}$ such that $\varphi_k \to 0$ and $T_\sigma \varphi_k \to f$ in $L^p(\mathbb{R}^n)$ as $k \to \infty$. Then for any function ψ in $\mathcal{S}$, we have

$$(T_\sigma \varphi_k, \psi) = (\varphi_k, T_\sigma^* \psi), \quad k = 1, 2, \ldots,$$

where T_σ^* is the formal adjoint of T_σ. Let $k \to \infty$. Then $(f, \psi) = 0$ for all functions $\psi \in \mathcal{S}$. Since $\mathcal{S}$ is dense in $L^p(\mathbb{R}^n)$, it follows that $f = 0$. Hence, by Definition 13.2, T_σ is closable. $\qquad\square$

Remark 14.2. A consequence of Proposition 14.1 is that the minimal operator $T_{\sigma,0}$ of T_σ exists. (See Remark 13.6.) Let us recall that the domain $\mathcal{D}(T_{\sigma,0})$ of $T_{\sigma,0}$ consists of all functions u in $L^p(\mathbb{R}^n)$ for which a sequence $\{\varphi_k\}$ in $\mathcal{S}$ can be found such that $\varphi_k \to u$ in $L^p(\mathbb{R}^n)$ and $T_\sigma \varphi_k \to f$ in $L^p(\mathbb{R}^n)$ for some $f \in L^p(\mathbb{R}^n)$ as $k \to \infty$. Moreover, $T_{\sigma,0} u = f$.

Definition 14.3. Let u and f be functions in $L^p(\mathbb{R}^n)$, $1 < p < \infty$. We say that u lies in $\mathcal{D}(T_{\sigma,1})$ and $T_{\sigma,1} u = f$ if and only if

$$(u, T_\sigma^* \varphi) = (f, \varphi), \quad \varphi \in \mathcal{S}, \tag{14.1}$$

where T_σ^* is the formal adjoint of T_σ.

Proposition 14.4. *Let* $u \in \mathcal{D}(T_{\sigma,1})$. *Then* $T_{\sigma,1}u = T_\sigma u$ *in the distribution sense.*

Proof By Definition 14.3,

$$(T_{\sigma,1}u, \varphi) = (u, T_\sigma^*\varphi), \quad \varphi \in \mathcal{S}.$$

Hence, by considering u and $T_{\sigma,1}u$ as tempered distributions, we have

$$(T_{\sigma,1}u)(\overline{\varphi}) = u(\overline{T_\sigma^*\varphi}), \quad \varphi \in \mathcal{S}. \tag{14.2}$$

On the other hand, by (11.3),

$$(T_\sigma u)(\overline{\varphi}) = u(\overline{T_\sigma^*\varphi}), \quad \varphi \in \mathcal{S}. \tag{14.3}$$

Hence, by (14.2) and (14.3), $T_{\sigma,1}u = T_\sigma u$ in the distribution sense. $\quad\square$

Proposition 14.5. $T_{\sigma,1}$ *is a closed linear operator from* $L^p(\mathbb{R}^n)$ *into* $L^p(\mathbb{R}^n)$ *with domain* $\mathcal{D}(T_{\sigma,1})$ *containing* $\mathcal{S}$.

Proof That $\mathcal{S} \subseteq \mathcal{D}(T_{\sigma,1})$ is obvious from (14.1) and the definition of the formal adjoint T_σ^*. Linearity is again easy to check. To prove that $T_{\sigma,1}$ is closed, let $\{u_k\}$ be a sequence of functions in $\mathcal{D}(T_{\sigma,1})$ such that $u_k \to u$ in $L^p(\mathbb{R}^n)$ and $T_{\sigma,1}u_k \to f$ in $L^p(\mathbb{R}^n)$ for some u and $f \in L^p(\mathbb{R}^n)$ as $k \to \infty$. Then, by (14.1),

$$(u_k, T_\sigma^*\varphi) = (T_{\sigma,1}u_k, \varphi) \tag{14.4}$$

for all $\varphi \in \mathcal{S}$ and $k = 1, 2, \ldots$. Let $k \to \infty$ in (14.4). Then

$$(u, T_\sigma^*\varphi) = (f, \varphi), \quad \varphi \in \mathcal{S}.$$

Hence, by Definition 14.3, $u \in \mathcal{D}(T_{\sigma,1})$ and $T_{\sigma,1}u = f$. This proves that $T_{\sigma,1}$ is closed by Definition 13.1. $\quad\square$

Proposition 14.6. $\mathcal{S} \subseteq \mathcal{D}(T_{\sigma,1}^t)$, *where* $T_{\sigma,1}^t$ *is the adjoint of* $T_{\sigma,1}$.

Proof Since $T_{\sigma,1}$ is a closed linear operator from $L^p(\mathbb{R}^n)$ into $L^p(\mathbb{R}^n)$, for $1 < p < \infty$, with domain containing $\mathcal{S}$, it follows from Proposition 13.8 that $T_{\sigma,1}^t$ is a closed linear operator from $L^{p'}(\mathbb{R}^n)$ into $L^{p'}(\mathbb{R}^n)$, where p' is the conjugate index of p. Let $\psi \in \mathcal{S}$. Then for all functions $u \in \mathcal{D}(T_{\sigma,1})$,

$$(\psi, T_{\sigma,1}u) = (T_\sigma^*\psi, u)$$

by Definition 14.3. Hence, by the definition of $T_{\sigma,1}^t$, $\psi \in \mathcal{D}(T_{\sigma,1}^t)$ and $T_{\sigma,1}^t\psi = T_\sigma^*\psi$. $\quad\square$

Proposition 14.7. $T_{\sigma,1}$ *is an extension of* $T_{\sigma,0}$.

Proof Let $u \in \mathcal{D}(T_{\sigma,0})$ and $T_{\sigma,0}u = f$. Then, by Remark 14.2, there is a sequence $\{\varphi_k\}$ of functions in $\mathcal{S}$ for which $\varphi_k \to u$ and $T_\sigma \varphi_k \to f$ in $L^p(\mathbb{R}^n)$ as $k \to \infty$. Hence, by the definition of T_σ^*, we have

$$(\varphi_k, T_\sigma^* \psi) = (T_\sigma \varphi_k, \psi)$$

for all $\psi \in \mathcal{S}$ and $k = 1, 2, \dots$. Let $k \to \infty$. Then

$$(u, T_\sigma^* \psi) = (f, \psi), \quad \psi \in \mathcal{S}.$$

So, by Definition 14.3, $u \in \mathcal{D}(T_{\sigma,1})$ and $T_{\sigma,1}u = f$. $\qquad\square$

Remark 14.8. Using Propositions 13.9 and 14.7, we see that $T_{\sigma,0}^t$ is an extension of $T_{\sigma,1}^t$. Since, by Proposition 14.6, the domain of $T_{\sigma,1}^t$ contains the space $\mathcal{S}$, it follows that the domain of $T_{\sigma,0}^t$ contains $\mathcal{S}$ as well.

Proposition 14.9. $T_{\sigma,1}$ *is the largest closed extension of* T_σ *in the sense that if* B *is any extension of* T_σ *such that* $\mathcal{S} \subseteq \mathcal{D}(B^t)$, *then* $T_{\sigma,1}$ *is an extension of* B.

We first prove the following lemma.

Lemma 14.10. $T_\sigma^t \varphi = T_\sigma^* \varphi$ *for all* $\varphi \in \mathcal{S}$. *In other words, the true and formal adjoints coincide on the space* $\mathcal{S}$.

Proof Let $\varphi \in \mathcal{S}$. Then, by the definition of T_σ^*, we have

$$(T_\sigma^* \varphi, \psi) = (\varphi, T_\sigma \psi), \quad \psi \in \mathcal{S}.$$

So, by the definition of T_σ^t and the duality of $L^p(\mathbb{R}^n)$, $\varphi \in \mathcal{D}(T_\sigma^t)$ and $T_\sigma^t \varphi = T_\sigma^* \varphi$. $\qquad\square$

Proof of Proposition 14.9 Let $u \in \mathcal{D}(B)$. Then for all $\psi \in \mathcal{S}$, we have $\psi \in \mathcal{D}(B^t)$. Hence, by the definition of B^t,

$$(\psi, Bu) = (B^t \psi, u). \tag{14.5}$$

Since B is an extension of T_σ, it follows from Proposition 13.9 that T_σ^t is an extension of B^t. Hence, by (14.5),

$$(\psi, Bu) = (T_\sigma^t \psi, u). \tag{14.6}$$

By Lemma 14.10, $T_\sigma^t = T_\sigma^*$ on $\mathcal{S}$. Hence, by (14.6), we have

$$(\psi, Bu) = (T_\sigma^* \psi, u), \quad \psi \in \mathcal{S}.$$

Therefore, by Definition 14.3, we have $u \in \mathcal{D}(T_{\sigma,1})$ and $T_{\sigma,1}u = Bu$. $\qquad\square$

Remark 14.11. Because of Proposition 14.9, we call $T_{\sigma,1}$ the *maximal operator* of T_σ.

By Proposition 14.7, we know that $T_{\sigma,1}$ is an extension of $T_{\sigma,0}$. The aim of this chapter is to prove that $T_{\sigma,0} = T_{\sigma,1}$ if σ is an elliptic symbol in $S^m, m > 0$. (See Chapter 10 for the definition of ellipticity.) We need some preparation.

Theorem 14.12. *Let $m > 0$ and σ be an elliptic symbol in S^m. Then $\mathcal{D}(T_{\sigma,0}) = H^{m,p}$.*

To prove Theorem 14.12, we use the following estimate, which is the analog of the Agmon–Douglas–Nirenberg estimate in [Agmon, Douglas and Nirenberg (1959)] for pseudo-differential operators.

Proposition 14.13. *Let $m > 0$ and σ be an elliptic symbol in S^m. Then there exist positive constants C_1 and C_2 such that*

$$C_1 \|u\|_{m,p} \leq (\|T_\sigma u\|_{0,p} + \|u\|_{0,p}) \leq C_2 \|u\|_{m,p}, \quad u \in H^{m,p}.$$

Proof By Theorems 12.5 and 12.9, there is a positive constant C' such that

$$\|T_\sigma u\|_{0,p} + \|u\|_{0,p} \leq C' \|u\|_{m,p}, \quad u \in H^{m,p}.$$

Next, by (10.1) in Theorem 10.1, we have

$$u = T_\tau T_\sigma u - Ru, \quad u \in H^{m,p}, \tag{14.7}$$

where $\tau \in S^{-m}$ and R is a pseudo-differential operator with symbol in $\cap_{k \in \mathbb{R}} S^k$. Hence it follows from Theorem 12.9 and (14.7) that there is a positive constant C such that

$$C\|u\|_{m,p} \leq (\|T_\sigma u\|_{0,p} + \|u\|_{0,p}), \quad u \in H^{m,p}.$$

This proves Proposition 14.13. $\square$

Proposition 14.14. *$\mathcal{S}$ is dense in $H^{s,p}$, $-\infty < s < \infty$, $1 < p < \infty$.*

Proof Let $u \in H^{s,p}$. Then, by the definition of $H^{s,p}$, $J_{-s}u \in L^p(\mathbb{R}^n)$. Since $\mathcal{S}$ is dense in $L^p(\mathbb{R}^n)$ by Remark 3.10, it follows that there is a sequence $\{\varphi_k\}$ of functions in $\mathcal{S}$ such that $\varphi_k \to J_{-s}u$ in $L^p(\mathbb{R}^n)$ as $k \to \infty$. Let $\psi_k = J_s \varphi_k$, $k = 1, 2, \ldots$ By Proposition 6.7, $\psi_k \in \mathcal{S}$, $k = 1, 2, \ldots$ Also, by the definition of $H^{s,p}$ again,

$$\|\psi_k - u\|_{s,p} = \|J_{-s}\psi_k - J_{-s}u\|_p = \|\varphi_k - J_{-s}u\|_p \to 0$$

as $k \to \infty$. This proves that $\mathcal{S}$ is dense in $H^{s,p}$. $\square$

We can now prove Theorem 14.12.

Proof of Theorem 14.12 Let $u \in H^{m,p}$. Then, by Proposition 14.14, we can find a sequence $\{\varphi_k\}$ of functions in $\mathcal{S}$ such that $\varphi_k \to u$ in $H^{m,p}$ as $k \to \infty$. By Propositions 14.13 and 14.14, $\{T_\sigma \varphi_k\}$ and $\{\varphi_k\}$ are Cauchy sequences in $L^p(\mathbb{R}^n)$. Hence $\varphi_k \to u$ and $T_\sigma \varphi_k \to f$ in $L^p(\mathbb{R}^n)$ for some u and f in $L^p(\mathbb{R}^n)$ as $k \to \infty$. Hence, by the definition of $T_{\sigma,0}$, $u \in \mathcal{D}(T_{\sigma,0})$ and $T_{\sigma,0}u = f$. On the other hand, if $u \in \mathcal{D}(T_{\sigma,0})$, then, by the definition of $T_{\sigma,0}$ again, we can find a sequence $\{\varphi_k\}$ of functions in $\mathcal{S}$ for which $\varphi_k \to u$ in $L^p(\mathbb{R}^n)$ and $T_\sigma \varphi_k \to f$ in $L^p(\mathbb{R}^n)$ for some $f \in L^p(\mathbb{R}^n)$ as $k \to \infty$. Hence $\{\varphi_k\}$ and $\{T_\sigma \varphi_k\}$ are Cauchy sequences in $L^p(\mathbb{R}^n)$. So, by Propositions 14.13 and 14.14, $\{\varphi_k\}$ is a Cauchy sequence in $H^{m,p}$. Since $H^{m,p}$ is complete by Theorem 12.3, it follows that $\varphi_k \to v$ in $H^{m,p}$ for some $v \in H^{m,p}$ as $k \to \infty$. Then, by Theorem 12.5, $\varphi_k \to v$ in $L^p(\mathbb{R}^n)$ as $k \to \infty$. Hence $u = v$ and consequently $u \in H^{m,p}$. $\qquad\square$

Finally we come to the main result of this chapter.

Theorem 14.15. *Let $m > 0$ and σ be an elliptic symbol in S^m. Then $T_{\sigma,0} = T_{\sigma,1}$.*

Proof Since $T_{\sigma,0}$ is the smallest closed extension of T_σ, it follows from Proposition 14.7 and Theorem 14.12 that it is sufficient to prove that $\mathcal{D}(T_{\sigma,1}) \subseteq H^{m,p}$. Let $u \in \mathcal{D}(T_{\sigma,1})$. Then, by (10.1) in Theorem 10.1,

$$u = T_\tau T_\sigma u - Ru, \qquad (14.8)$$

where $\tau \in S^{-m}$ and R is a pseudo-differential operator with symbol in $\cap_{k \in \mathbb{R}} S^k$. By Proposition 14.4, $T_{\sigma,1}u = T_\sigma u$ in the distribution sense. Thus, by the definition of $T_{\sigma,1}$, $T_\sigma u \in L^p(\mathbb{R}^n)$. Since $\tau \in S^{-m}$, it follows from Theorem 12.9 that $T_\tau T_\sigma u \in H^{m,p}$. Since $u \in L^p(\mathbb{R}^n)$ and R has symbol in S^{-m}, it follows from Theorem 12.9 again that $Ru \in H^{m,p}$. Hence, by (14.8), $u \in H^{m,p}$. $\qquad\square$

Exercises

14.1. Let σ be any symbol in S^m, $m \leq 0$. Prove that the minimal operator $T_{\sigma,0}$ of $T_\sigma : \mathcal{S} \to \mathcal{S}$ in $L^p(\mathbb{R}^n)$, $1 < p < \infty$, is a bounded linear operator from $L^p(\mathbb{R}^n)$ into $L^p(\mathbb{R}^n)$.

14.2. Let σ be any symbol. Consider T_σ as a linear operator from $L^p(\mathbb{R}^n)$ into $L^p(\mathbb{R}^n)$, $1 < p < \infty$, with dense domain $\mathcal{S}$. Prove that $T_\sigma^t = (T_{\sigma,0})^t$.

14.3. Let σ and T_σ be as in Exercise 14.2. Prove that $T_\sigma^t = (T_\sigma^*)_1$, where $(T_\sigma^*)_1$ is the maximal operator of the pseudo-differential operator T_σ^*.

14.4. For any closed linear operator A from a complex Banach space into itself with domain $\mathcal{D}(A)$, a subspace $\mathcal{D}$ of $\mathcal{D}(A)$ is called a *core* of the operator A if the minimal operator of the restriction of A to $\mathcal{D}$ is equal to A. Prove that if σ is any elliptic symbol, then $C_0^\infty(\mathbb{R}^n)$ is a core of $T_{\sigma,0}$ and $(T_{\sigma,0})^t$.

Chapter 15

Global Regularity of Elliptic Partial Differential Equations

Let $P(x, D) = \sum_{|\alpha| \leq m} a_\alpha(x) D^\alpha$ be a linear partial differential operator of order m such that

$$\sup_{x \in \mathbb{R}^n} |(D^\beta a_\alpha)(x)| < \infty, \quad |\alpha| \leq m, \tag{15.1}$$

for all multi-indices β. Then we have observed in Example 6.3 that $P(x, D)$ is a pseudo-differential operator with symbol $P(x, \xi)$ in S^m, where

$$P(x, \xi) = \sum_{|\alpha| \leq m} a_\alpha(x) \xi^\alpha.$$

The purpose of this chapter is to use the theory of pseudo-differential operators we have developed to prove the following L^p analog of a result in [Hess and Kato (1970)].

Theorem 15.1. *Let $P(x, D) = \sum_{|\alpha| \leq m} a_\alpha(x) D^\alpha$ be a linear partial differential operator of order m satisfying (15.1). Suppose that there exists a point $x_0 \in \mathbb{R}^n$ such that we can find positive constants C_1 and C_2 for which*

$$\left| \sum_{|\alpha|=m} a_\alpha(x_0) \xi^\alpha \right| \geq C_1 |\xi|^m \tag{15.2}$$

and

$$\left| \sum_{|\alpha|=m} (a_\alpha(x) - a_\alpha(x_0)) \xi^\alpha \right| \leq C_2 |\xi|^m \tag{15.3}$$

for all x and $\xi \in \mathbb{R}^n$, and $C_2 < C_1$. If $u \in H^{s,p}$, $P(x, D)u = f$ and $f \in H^{s,p}$, then $u \in H^{s+m,p}$.

Remark 15.2. If $P(x, D)$ is a linear partial differential operator of order m satisfying the hypotheses of Theroem 15.1, and if f is any tempered distribution in $H^{s,p}$, then Theorem 15.1 asserts that any solution u in $H^{s,p}$ of the partial differential equation $P(x, D)u = f$ on $\mathbb{R}^n$ lies in a more "selective" or "regular" space $H^{s+m,p}$. If we recall, by Theorem 12.5, that

$$H^{s+m,p} \subset H^{s+m-1,p} \subset \cdots \subset H^{s,p},$$

then the solution u can be thought of being m steps more "selective" or "regular" than the given source data f defined globally on $\mathbb{R}^n$. For this reason, we call Theorem 15.1 a global regularity theorem.

To prove Theorem 15.1, we use the following lemma.

Lemma 15.3. *Let $P(x, D) = \sum_{|\alpha| \le m} a_\alpha(x)D^\alpha$ be a linear partial differential operator of order m satisfying the hypotheses of Theorem 15.1. Then there exist positive constants C and R such that*

$$|P(x, \xi)| \ge C(1 + |\xi|)^m, \quad |\xi| \ge R.$$

Remark 15.4. Lemma 15.3 tells us that under the hypotheses of Theorem 15.1, $P(x, D)$ is a pseudo-differential operator with symbol $P(x, \xi)$ in S^m and $P(x, \xi)$ satisfies the ellipticity condition defined in Chapter 10.

Proof of Lemma 15.3 By (15.2) and (15.3), we obtain

$$
\left| \sum_{|\alpha|=m} a_\alpha(x)\xi^\alpha \right|
$$

$$
= \left| \sum_{|\alpha|=m} (a_\alpha(x) - a_\alpha(x_0))\xi^\alpha + \sum_{|\alpha|=m} a_\alpha(x_0)\xi^\alpha \right|
$$

$$
\ge \left| \sum_{|\alpha|=m} a_\alpha(x_0)\xi^\alpha \right| - \left| \sum_{|\alpha|=m} (a_\alpha(x) - a_\alpha(x_0))\xi^\alpha \right|
$$

$$
\ge (C_1 - C_2)|\xi|^m, \quad x, \xi \in \mathbb{R}^n. \tag{15.4}
$$

Next, by (15.4), we can find positive constants C', C'' and R such that

$$\left| \sum_{|\alpha| \leq m} a_\alpha(x) \xi^\alpha \right|$$

$$= \left| \sum_{|\alpha| = m} a_\alpha(x) \xi^\alpha + \sum_{|\alpha| < m} a_\alpha(x) \xi^\alpha \right|$$

$$\geq \left| \sum_{|\alpha| = m} a_\alpha(x) \xi^\alpha \right| - \left| \sum_{|\alpha| < m} a_\alpha(x) \xi^\alpha \right|$$

$$\geq C'(1 + |\xi|)^m - C''(1 + |\xi|)^{m-1}$$

$$= (1 + |\xi|)^m (C' - C''(1 + |\xi|)^{-1}), \quad |\xi| \geq R. \tag{15.5}$$

Obviously, we can find another positive constant $R_1 \geq R$ such that

$$C' - C''(1 + |\xi|)^{-1} \geq \frac{C'}{2}, \quad |\xi| \geq R_1. \tag{15.6}$$

Hence, by (15.5) and (15.6),

$$\left| \sum_{|\alpha| \leq m} a_\alpha(x) \xi^\alpha \right| \geq \frac{C'}{2} (1 + |\xi|)^m, \quad |\xi| \geq R_1.$$

This proves Lemma 15.3. $\qquad\qquad\qquad\qquad\qquad\qquad\qquad\qquad\quad\square$

Proof of Theorem 15.1 By Lemma 15.3, $P(x, D)$ is an elliptic pseudo-differential operator with symbol $P(x, \xi)$ in S^m. Hence, by Theorem 10.1, we can find a symbol $\tau \in S^{-m}$ and a pseudo-differential operator R with symbol in $\cap_{k \in \mathbb{R}} S^k$ such that

$$T_\tau P(x, D) = I + R. \tag{15.7}$$

Hence, by (15.7), $u = T_\tau f - Ru$. Since $\tau \in S^{-m}$ and $f \in H^{s,p}$, it follows that $T_\tau f \in H^{s+m,p}$. Also, $Ru \in H^{s+m,p}$ because R has symbol in S^{-m} and $u \in H^{s,p}$. Hence, by (15.7), $u \in H^{s+m,p}$. $\qquad\qquad\square$

Exercises

15.1. Let $P(x, D)$ be a linear partial differential operator of order m satisfying all the hypotheses of Theorem 15.1. Prove that if $f \in H^{s,p}$, then any solution u in $\cup_{t \in \mathbb{R}} H^{t,p}$ of the partial differential equation $P(x, D)u = f$ on $\mathbb{R}^n$ is in $H^{s+m,p}$.

15.2. Let $m > 0$ and $\sigma \in S^m$ be an elliptic symbol. Let $f \in L^p(\mathbb{R}^n)$, for $1 < p < \infty$ and u be any solution in $\cup_{t \in \mathbb{R}} H^{t,p}$ of the pseudo-differential equation $T_\sigma u = f$ on $\mathbb{R}^n$.

(i) Prove that $u \in L^p(\mathbb{R}^n)$.

(ii) Prove that $(u, T_\sigma^* \varphi) = (f, \varphi)$ for all $\varphi \in \mathcal{S}$.

(iii) Prove that there exists a sequence $\{\varphi_k\}$ of functions in $\mathcal{S}$ such that $\varphi_k \to u$ in $L^p(\mathbb{R}^n)$ and $T_\sigma \varphi_k \to f$ in $L^p(\mathbb{R}^n)$ as $k \to \infty$.

Chapter 16

Weak Solutions of Pseudo-Differential Equations

We give in this chapter a result on the existence of weak solutions in $L^p(\mathbb{R}^n)$, $1 < p < \infty$, of pseudo-differential equations on $\mathbb{R}^n$. We begin with the definition of a weak solution.

Definition 16.1. Let $\sigma \in S^m$, $m > 0$, and $f \in L^p(\mathbb{R}^n)$, $1 < p < \infty$. A function u in $L^p(\mathbb{R}^n)$ is said to be a *weak solution* of the pseudo-differential equation $T_\sigma u = f$ on $\mathbb{R}^n$ if

$$(u, T_\sigma^* \varphi) = (f, \varphi), \quad \varphi \in \mathcal{S},$$

where T_σ^* is the formal adjoint of T_σ introduced in Chapter 9.

From the definition of the maximal operator $T_{\sigma,1}$ of T_σ given in Chapter 14, it is obvious that the following proposition is true.

Proposition 16.2. *Let $\sigma \in S^m$, $m > 0$, and $f \in L^p(\mathbb{R}^n)$, $1 < p < \infty$. Then a function u in $L^p(\mathbb{R}^n)$ is a weak solution of the pseudo-differential equation $T_\sigma u = f$ on $\mathbb{R}^n$ if and only if $u \in \mathcal{D}(T_{\sigma,1})$ and $T_{\sigma,1}u = f$.*

For any $\sigma \in S^m$, $m > 0$, the following theorem characterizes the functions f in $L^p(\mathbb{R}^n)$, $1 < p < \infty$, for which the pseudo-differential equation $T_\sigma u = f$ on $\mathbb{R}^n$ has a weak solution u in $L^p(\mathbb{R}^n)$.

Theorem 16.3. *Let $\sigma \in S^m$, $m > 0$, and $f \in L^p(\mathbb{R}^n)$, $1 < p < \infty$. Then the pseudo-differential equation $T_\sigma u = f$ on $\mathbb{R}^n$ has a weak solution u in $L^p(\mathbb{R}^n)$ if and only if there exists a positive constant C such that*

$$|(f, \varphi)| \leq C \|T_\sigma^* \varphi\|_{p'}, \quad \varphi \in \mathcal{S}, \tag{16.1}$$

where p' is the conjugate index of p.

Proof Suppose that $T_\sigma u = f$ on $\mathbb{R}^n$ has a weak solution u in $L^p(\mathbb{R}^n)$. Then, by Definition 16.1,

$$(f, \varphi) = (u, T_\sigma^* \varphi), \quad \varphi \in \mathcal{S}.$$

Hence, by Hölder's inequality,

$$|(f, \varphi)| \le \|u\|_p \|T_\sigma^* \varphi\|_{p'}, \quad \varphi \in \mathcal{S},$$

and the inequality (16.1) holds with $C = \|u\|_p$. Conversely, suppose that the inequality (16.1) is true. Let W be the subspace of $L^{p'}(\mathbb{R}^n)$ defined by

$$W = \{T_\sigma^* \varphi : \varphi \in \mathcal{S}\}.$$

We define the linear functional $F : W \to \mathbb{C}$ by

$$Fw = (\varphi, f), \quad w \in W,$$

where φ is any function in $\mathcal{S}$ with the property that $T_\sigma^* \varphi = w$. To see that the defintion of $F : W \to \mathbb{C}$ is independent of the function φ, let φ_1 and φ_2 be functions in $\mathcal{S}$ such that $T_\sigma^* \varphi_1 = w$ and $T_\sigma^* \varphi_2 = w$. Then, by (16.1),

$$|(\varphi_1 - \varphi_2, f)| \le C \|T_\sigma^*(\varphi_1 - \varphi_2)\|_{p'} = 0.$$

Hence $(\varphi_1, f) = (\varphi_2, f)$ and this proves that the choice of the function φ is irrelevant to the definition of $F : W \to \mathbb{C}$. Since, by (16.1),

$$|Fw| = |(\varphi, f)| \le C \|T_\sigma^* \varphi\|_{p'} = C \|w\|_{p'}, \quad w \in W,$$

it follows that $F : W \to \mathbb{C}$ is a bounded linear functional. Hence, using the Hahn–Banach theorem and the Riesz representation theorem, we can find a function u in $L^p(\mathbb{R}^n)$ such that

$$Fw = (\varphi, f) = (w, u), \quad w \in W, \tag{16.2}$$

where φ is any function in $\mathcal{S}$ satisfying $T_\sigma^* \varphi = w$. Since $\{T_\sigma^* \varphi : \varphi \in \mathcal{S}\}$ is obviously a subspace of W, it follows from (16.2) that

$$(\varphi, f) = (T_\sigma^* \varphi, u), \quad \varphi \in \mathcal{S},$$

and hence, by Definition 16.1, u is a weak solution in $L^p(\mathbb{R}^n)$ of the pseudo-differential equation $T_\sigma u = f$ on $\mathbb{R}^n$. $\qquad \square$

Exercises

16.1. Let $\sigma \in S^m$, $m > 0$, and let u and f be in $L^p(\mathbb{R}^n)$, $1 < p < \infty$. Prove that u is a solution of $T_\sigma u = f$ on $\mathbb{R}^n$ in the distribution sense if and only if u is a weak solution of $T_\sigma u = f$ on $\mathbb{R}^n$.

16.2. Let $\sigma \in S^m$, $m > 0$, be an elliptic symbol, and let $f \in L^p(\mathbb{R}^n)$, for $1 < p < \infty$. Prove that every weak solution u in $L^p(\mathbb{R}^n)$ of the pseudo-differential equation $T_\sigma u = f$ on $\mathbb{R}^n$ is in $H^{m,p}$.

16.3. Let $s > 0$ and J_{-s} be the pseudo-differential operator defined in Chapter 12. Let q be any real-valued and nonnegative function on $\mathbb{R}^n$ such that

$$\sup_{x \in \mathbb{R}^n} |(D^\alpha q)(x)| < \infty$$

for all multi-indices α. Prove that the pseudo-differential equation

$$J_{-s}u + qu = f$$

on $\mathbb{R}^n$ has a weak solution u in $L^2(\mathbb{R}^n)$ for every function f in $L^2(\mathbb{R}^n)$.

16.4. Let $\sigma(x, \xi) \in S^m$, $m > 0$, be such that σ is independent of $x \in \mathbb{R}^n$. Prove that the pseudo-differential equation $T_\sigma u = f$ on $\mathbb{R}^n$ has a unique weak solution u in $L^2(\mathbb{R}^n)$ for all functions f in $L^2(\mathbb{R}^n)$ if and only if there exists a positive constant C such that

$$|\sigma(\xi)| \geq C, \quad \xi \in \mathbb{R}^n.$$

Chapter 17

Gårding's Inequality

We are interested in a subset of the set of all elliptic pseudo-differential operators introduced in Chapter 10. These operators satisfy an important inequality in the study of pseudo-differential operators.

Theorem 17.1. (Gårding's Inequality) *Let $\sigma \in S^{2m}$ be such that there exist positive constants C and R for which*

$$\operatorname{Re}\sigma(x,\xi) \geq C(1 + |\xi|)^{2m}, \quad |\xi| \geq R.$$

Then we can find a positive constant C' and a constant C_s for every real number $s \geq \frac{1}{2}$ such that

$$\operatorname{Re}(T_\sigma\varphi, \varphi) \geq C'\|\varphi\|_{m,2}^2 - C_s\|\varphi\|_{m-s,2}^2, \quad \varphi \in \mathcal{S}.$$

A symbol satisfying the hypothesis of the theorem is said to be *strongly elliptic*. In order to prove the theorem, we need two lemmas.

Lemma 17.2. *Let F be a C^∞ function on the complex plane $\mathbb{C}$. Then for every σ in S^0, $F \circ \sigma \in S^0$.*

Proof We need to prove that for all multi-indices α and β, there exists a positive constant $C_{\alpha,\beta}$ such that

$$|(\partial_x^\alpha \partial_\xi^\beta (F \circ \sigma))(x,\xi)| \leq C_{\alpha,\beta}(1 + |\xi|)^{-|\beta|}, \quad x, \xi \in \mathbb{R}^n. \tag{17.1}$$

(17.1) is true for all multi-indices α and β with $|\alpha + \beta| = 0$. Indeed, there exists a positive constant C such that

$$|\sigma(x,\xi)| \leq C, \quad x, \xi \in \mathbb{R}^n.$$

Thus, $F \circ \sigma$ is in fact a bounded and C^∞ function on $\mathbb{R}^n \times \mathbb{R}^n$. Hence there exists another positive constant C' such that

$$|(F \circ \sigma)(x,\xi)| \leq C', \quad x, \xi \in \mathbb{R}^n.$$

Now, suppose that (17.1) is valid for all C^∞ functions F on $\mathbb{C}$, σ in S^0 and multi-indices α and β with $|\alpha + \beta| = l$. Let α and β be multi-indices with $|\alpha + \beta| = l + 1$. We first suppose that

$$\partial_x^\alpha \partial_\xi^\beta = \partial_x^\alpha \partial_\xi^\gamma \partial_{\xi_j}$$

for some multi-index γ and some $j = 1, 2, \ldots, n$. Then, by the chain rule,

$$(\partial_x^\alpha \partial_\xi^\beta (F \circ \sigma))(x, \xi) = \partial_x^\alpha \partial_\xi^\gamma \{(F_1 \circ \sigma)\partial_{\xi_j}\sigma + (F_2 \circ \sigma)\partial_{\xi_j}\sigma\}(x, \xi)$$

for all x and ξ in $\mathbb{R}^n$, where F_1 and F_2 are the partial derivatives of F with respect to the first and second variables respectively. Now, by Leibniz's formula and the induction hypothesis, there exist positive constants $C_{\rho,\delta}$ and $C_{\alpha,\rho,\gamma,\delta,j}$ such that

$$|(\partial_x^\alpha \partial_\xi^\gamma \{(F_1 \circ \sigma)\partial_{\xi_j}\sigma\})(x, \xi)|$$

$$\leq \sum_{\substack{\rho \leq \alpha \\ \delta \leq \gamma}} \binom{\alpha}{\rho}\binom{\gamma}{\delta} |(\partial_x^\rho \partial_\xi^\delta (F_1 \circ \sigma))(x, \xi)| \, |(\partial_x^{\alpha-\rho} \partial_\xi^{\gamma-\delta}(\partial_{\xi_j}\sigma))(x, \xi)|$$

$$\leq \sum_{\substack{\rho \leq \alpha \\ \delta \leq \gamma}} \binom{\alpha}{\rho}\binom{\gamma}{\delta} C_{\rho,\delta}(1 + |\xi|)^{-|\delta|} C_{\alpha,\rho,\gamma,\delta,j}(1 + |\xi|)^{-|\gamma|+|\delta|-1}$$

$$= C_{\alpha,\gamma,j}(1 + |\xi|)^{-(|\gamma|+1)}, \quad x, \xi \in \mathbb{R}^n,$$

where

$$C_{\alpha,\gamma,j} = \sum_{\substack{\rho \leq \alpha \\ \delta \leq \gamma}} \binom{\alpha}{\rho}\binom{\gamma}{\delta} C_{\rho,\delta} C_{\alpha,\rho,\gamma,\delta,j}.$$

Similarly, there exists a positive constant $C'_{\alpha,\gamma,j}$ such that

$$|(\partial_x^\alpha \partial_\xi^\gamma \{(F_2 \circ \sigma)\partial_{\xi_j}\sigma\}(x, \xi)| \leq C'_{\alpha,\gamma,j}(1 + |\xi|)^{-(|\gamma|+1)},$$

for all x and ξ in $\mathbb{R}^n$. Therefore

$$|(\partial_x^\alpha \partial_\xi^\beta (F \circ \sigma))(x, \xi)| \leq (C_{\alpha,\gamma,j} + C'_{\alpha,\gamma,j})(1 + |\xi|)^{-|\beta|}$$

for all x and ξ in $\mathbb{R}^n$. Now, we suppose that

$$\partial_x^\alpha \partial_\xi^\beta = \partial_x^\gamma \partial_{x_j} \partial_\xi^\beta$$

for some multi-index γ and some $j = 1, 2, \ldots, n$. Then, as before, there exists a positive constant $C''_{\alpha,\gamma,j}$ such that

$$|(\partial_x^\alpha \partial_\xi^\beta (F \circ \sigma))(x, \xi)| \leq C''_{\alpha,\gamma,j}(1 + |\xi|)^{-|\beta|}$$

for all x and ξ in $\mathbb{R}^n$. Thus, by the principle of mathematical induction, (17.1) follows. $\qquad\square$

Lemma 17.3. *Let σ be a strongly elliptic symbol in S^{2m}, $m > 0$. Then there exist positive constants γ and κ such that*

$$\operatorname{Re}\sigma(x,\xi) \geq \gamma\langle\xi\rangle^{2m} - \kappa\langle\xi\rangle^{2m-1}, \quad x,\xi \in \mathbb{R}^n,$$

where $\langle\ \rangle$ is the function on $\mathbb{R}^n$ defined by

$$\langle\xi\rangle = (1 + |\xi|^2)^{1/2}, \quad \xi \in \mathbb{R}^n.$$

Proof By strong ellipticity, there exist positive constants C and R such that

$$\operatorname{Re}\sigma(x,\xi) \geq C\langle\xi\rangle^{2m}, \quad |\xi| \geq R.$$

Since $\sigma \in S^{2m}$, we can find a positive constant K such that

$$|\sigma(x,\xi)| \leq K\langle\xi\rangle^{2m}, \quad x,\xi \in \mathbb{R}^n.$$

Therefore

$$|\operatorname{Re}\sigma(x,\xi)| \leq K\langle\xi\rangle^{2m} \leq K(1+R^2)^m, \quad |\xi| \leq R.$$

Hence there exists a positive constant M such that

$$\operatorname{Re}\sigma(x,\xi) \geq -M, \quad |\xi| \leq R.$$

Since $\frac{\operatorname{Re}\sigma}{\langle\ \rangle^{2m-1}}$ is continuous on the compact set $\{\xi \in \mathbb{R}^n : |\xi| \leq R\}$, we can find a positive constant κ such that

$$\frac{\operatorname{Re}\sigma(x,\xi)}{\langle\xi\rangle^{2m-1}} > -\kappa, \quad |\xi| \leq R.$$

Therefore

$$\operatorname{Re}\sigma(x,\xi) + \kappa\langle\xi\rangle^{2m-1} > 0, \quad |\xi| \leq R.$$

Since $\frac{\operatorname{Re}\sigma + \kappa\langle\ \rangle^{2m-1}}{\langle\ \rangle^{2m}}$ is a positive and continuous function on the compact set $\{\xi \in \mathbb{R}^n : |\xi| \leq R\}$, there is a positive constant δ such that

$$\frac{\operatorname{Re}\sigma(x,\xi) + \kappa\langle\xi\rangle^{2m-1}}{\langle\xi\rangle^{2m}} \geq \delta, \quad |\xi| \leq R.$$

So, the lemma is proved if we let $\gamma = \min(C,\delta)$. $\qquad\square$

Proof of Theorem 17.1 Let $T_\tau = J_m T_\sigma J_m$, where J_m is the Bessel potential of order m. In fact, $J_m = T_{\langle\ \rangle^{-m}}$. Then, using the asymptotic expansion for the product of two pseudo-differential operators in Theorem 8.1,

$$T_\sigma J_m = T_{\tau_1},$$

where

$$\tau_1 - \langle \ \rangle^{-m} \sigma \in S^{m-1}. \tag{17.2}$$

Similarly,

$$T_\tau = J_m T_{\tau_1}$$

and

$$\tau - \langle \ \rangle^{-m} \tau_1 \in S^{-1}. \tag{17.3}$$

Multiplying (17.2) by $\langle \ \rangle^{-m}$ and adding the result to (17.3), we get

$$\tau - \langle \ \rangle^{-2m} \sigma \in S^{-1}.$$

Therefore

$$\tau = \langle \ \rangle^{-2m} \sigma + r,$$

where $r \in S^{-1}$. So, by Lemma 17.3,

$$\mathrm{Re}\, \tau = \langle \ \rangle^{-2m} \mathrm{Re}\, \sigma + \mathrm{Re}\, r \geq \gamma - \kappa \langle \ \rangle^{-1} + \mathrm{Re}\, r \geq \gamma - \kappa' \langle \ \rangle^{-1},$$

where κ' is another positive constant. Therefore τ satisfies the conclusion of Lemma 17.3 with $m = 0$. Let us suppose for a moment that Gårding's inequality is valid for $m = 0$. Then we can find a positive constant C' and a positive constant C_s for every real number $s \geq \frac{1}{2}$ such that

$$\begin{aligned}
\mathrm{Re}\,(T_\sigma \varphi, \varphi) &= \mathrm{Re}\,(J_{-m} T_\tau J_{-m} \varphi, \varphi) \\
&= \mathrm{Re}\,(T_\tau J_{-m} \varphi, J_{-m} \varphi) \\
&\geq C' \|J_{-m} \varphi\|_{0,2}^2 - C_s \|J_{-m} \varphi\|_{-s,2}^2 \\
&= C' \|\varphi\|_{m,2}^2 - C_s \|\varphi\|_{m-s,2}^2
\end{aligned}$$

for all φ in $\mathcal{S}$. We are now ready to prove Gårding's inequality for $m = 0$. By Lemma 17.3, we have positive constants γ and κ such that

$$\mathrm{Re}\, \sigma + \kappa \langle \ \rangle^{-1} \geq \gamma.$$

Let F be a C^∞ function on $\mathbb{C}$ such that

$$F(z) = \sqrt{\frac{\gamma}{2} + z}, \quad z \in [0, \infty).$$

Let τ be the function defined on $\mathbb{R}^n \times \mathbb{R}^n$ by

$$\tau(x,\xi) = F(2(\text{Re}\,\sigma(x,\xi) + \kappa\langle\xi\rangle^{-1} - \gamma)), \quad x,\xi \in \mathbb{R}^n.$$

Then, by Lemma 17.2, $\tau \in S^0$, and for all x and ξ in $\mathbb{R}^n$,

$$\tau(x,\xi) = \sqrt{\frac{\gamma}{2} + 2\text{Re}\,\sigma(x,\xi) + 2\kappa\langle\xi\rangle^{-1} - 2\gamma}$$

$$= \sqrt{2\text{Re}\,\sigma(x,\xi) + 2\kappa\langle\xi\rangle^{-1} - \frac{3}{2}\gamma}.$$

Using the asymptotic expansion for the formal adjoint of a pseudo-differential operator in Theorem 9.1, we have

$$T_\tau^* = T_{\tau^*},$$

where $\tau^* \in S^0$ and $\tau - \tau^* \in S^{-1}$. Using also the asymptotic expansion for the product in Theorem 8.1,

$$T_\tau^* T_\tau = T_\lambda,$$

where

$$\lambda - \tau^*\tau \in S^{-1}.$$

If we let r_1 and r_1' in S^{-1} be such that

$$\tau^* = \tau + r_1$$

and

$$\lambda = \tau^*\tau + r_1',$$

then, with $r_2 = r_1\tau + r_1' \in S^{-1}$,

$$\lambda = (\tau + r_1)\tau + r_1' = 2\text{Re}\,\sigma + 2\kappa\langle\,\rangle^{-1} - \frac{3}{2}\gamma + r_2.$$

So, if we let $r_3 = 2\kappa\langle\,\rangle^{-1} + r_2 \in S^{-1}$, then we get

$$\lambda = 2\text{Re}\,\sigma - \frac{3}{2}\gamma + r_3.$$

But

$$2\text{Re}\,\sigma = \sigma + \overline{\sigma} = \sigma + \sigma^* + r_4$$

for some r_4 in S^{-1}. Therefore

$$\lambda = \sigma + \sigma^* - \frac{3}{2}\gamma + r_5$$

for some r_5 in S^{-1}. Thus,

$$\sigma + \sigma^* = \lambda + \frac{3}{2}\gamma - r_5.$$

Since
$$(T_\lambda \varphi, \varphi) = (T_\tau \varphi, T_\tau \varphi) \geq 0, \quad \varphi \in \mathcal{S},$$
it follows that
$$2\text{Re}\,(T_\sigma \varphi, \varphi) = (T_\sigma \varphi, \varphi) + (T_\sigma^* \varphi, \varphi) = (T_{\sigma + \sigma^*} \varphi, \varphi)$$
$$= (T_\lambda \varphi, \varphi) + \frac{3}{2}\gamma \|\varphi\|_{0,2}^2 - (T_{r_5} \varphi, \varphi)$$
$$\geq \gamma \|\varphi\|_{0,2}^2 + \left\{ \frac{\gamma}{2}\|\varphi\|_{0,2}^2 - \|T_{r_5}\varphi\|_{\frac{1}{2},2}\|\varphi\|_{-\frac{1}{2},2} \right\}.$$
Using the L^2-boundedness of pseudo-differential operators in Theorem 11.7, we get a positive constant μ such that
$$2\text{Re}\,(T_\sigma \varphi, \varphi) \geq \gamma \|\varphi\|_{0,2}^2 + \left\{ \frac{\gamma}{2}\|\varphi\|_{0,2}^2 - \mu\|\varphi\|_{-\frac{1}{2},2}^2 \right\}, \quad \varphi \in \mathcal{S}.$$
But
$$\mu\|\varphi\|_{-\frac{1}{2},2}^2 = \int_{\mathbb{R}^n} \mu\langle\xi\rangle^{-1}|\hat{\varphi}(\xi)|^2\,d\xi = I + J,$$
where
$$I = \int_{\mu\langle\xi\rangle^{-1} \leq \frac{\gamma}{2}} \mu\langle\xi\rangle^{-1}|\hat{\varphi}(\xi)|^2 d\xi$$
and
$$J = \int_{\mu\langle\xi\rangle^{-1} \geq \frac{\gamma}{2}} \mu\langle\xi\rangle^{-1}|\hat{\varphi}(\xi)|^2 d\xi.$$
Obviously,
$$I \leq \frac{\gamma}{2} \int_{\mathbb{R}^n} |\hat{\varphi}(\xi)|^2 d\xi = \frac{\gamma}{2}\|\varphi\|_{0,2}^2.$$
To estimate J, we note that
$$\mu\langle\xi\rangle^{-1} \geq \frac{\gamma}{2} \Rightarrow \langle\xi\rangle \leq \frac{2\mu}{\gamma}.$$
So, for $\mu\langle\xi\rangle^{-1} \geq \frac{\gamma}{2}$, we get, for every real number $s \geq \frac{1}{2}$,
$$\mu\langle\xi\rangle^{-1} = \mu\langle\xi\rangle^{2s-1}\langle\xi\rangle^{-2s}$$
$$\leq \mu \left(\frac{2\mu}{\gamma}\right)^{2s-1} \langle\xi\rangle^{-2s}.$$
Then for every real number $s \geq \frac{1}{2}$,
$$J \leq \mu \left(\frac{2\mu}{\gamma}\right)^{2s-1} \int_{\mathbb{R}^n} \langle\xi\rangle^{-2s}|\hat{\varphi}(\xi)|^2 d\xi = C_s'\|\varphi\|_{-s,2}^2,$$
where $C_s' = \mu \left(\frac{2\mu}{\gamma}\right)^{2s-1}$. Therefore
$$2\text{Re}\,(T_\sigma \varphi, \varphi) \geq \gamma \|\varphi\|_{0,2}^2 - C_s'\|\varphi\|_{-s,2}^2, \quad \varphi \in \mathcal{S},$$
and the proof of the theorem is complete. $\square$

Exercises

17.1. Prove that a strongly elliptic pseudo-differential operator is elliptic.

17.2. Give an example of an elliptic pseudo-differential operator that is not strongly elliptic.

17.3. Let $\sigma \in S^{2m}$, $m \geq \frac{1}{2}$, be such that σ is strongly elliptic. Prove that there exists a constant C such that

$$\operatorname{Re}(T_\sigma \varphi, \varphi) \geq -C\|\varphi\|_2^2, \quad \varphi \in \mathcal{S}.$$

Chapter 18

Strong Solutions of Pseudo-Differential Equations

Let $\sigma \in S^m$, $m > 0$, and let $f \in L^p(\mathbb{R}^n)$, $1 < p < \infty$. From Chapter 16 we see that a function $u \in L^p(\mathbb{R}^n)$ is a weak solution of the pseudo-differential equation $T_\sigma u = f$ on $\mathbb{R}^n$ if $u \in \mathcal{D}(T_{\sigma,1})$ and $T_{\sigma,1}u = f$. In this chapter we give another notion of a solution of the equation. A function $u \in L^p(\mathbb{R}^n)$ is said to be a *strong solution* of the equation $T_\sigma u = f$ if $u \in \mathcal{D}(T_{\sigma,0})$ and $T_{\sigma,0}u = f$.

Remark 18.1. For the pseudo-differential equation $T_\sigma u = f$ on $\mathbb{R}^n$, it should be clear that strong solutions are weak solutions. If σ is elliptic, then it is also true that weak solutions are strong solutions. These simple facts are best left as exercises. See Exercises 18.1 and 18.2.

We begin with strong solutions in $L^2(\mathbb{R}^n)$.

Theorem 18.2. *Let $\sigma \in S^{2m}$, $m > 0$, be an elliptic symbol such that there exists a positive constant C for which*

$$\mathrm{Re}\,(T_\sigma \varphi, \varphi) \geq C\|\varphi\|_{m,2}^2, \quad \varphi \in \mathcal{S}. \tag{18.1}$$

Then for every function f in $L^2(\mathbb{R}^n)$, the pseudo-differential equation $T_\sigma u = f$ on $\mathbb{R}^n$ has a unique strong solution u in $L^2(\mathbb{R}^n)$.

We need the following lemma.

Lemma 18.3. *Under the hypotheses of the preceding theorem, there exists a positive constant C such that*

$$\mathrm{Re}\,(T_\sigma u, u) \geq C\|u\|_{m,2}^2, \quad u \in H^{m,2}.$$

Proof Let $u \in H^{m,2}$. Then, by Proposition 14.14, there exists a sequence $\{\varphi_k\}$ of functions in $\mathcal{S}$ such that $\varphi_k \to u$ in $H^{m,2}$ as $k \to \infty$. By Theorem

121

12.9, $T_\sigma : H^{m,2} \to H^{-m,2}$ is a bounded linear operator. Hence, by Exercise 12.14, there exists a positive constant C' such that

$$
\begin{aligned}
&|(T_\sigma \varphi_k, \varphi_k) - (T_\sigma u, u)| \\
&= |(T_\sigma \varphi_k, \varphi_k) - (T_\sigma u, \varphi_k) + (T_\sigma u, \varphi_k) - (T_\sigma u, u)| \\
&= |(T_\sigma(\varphi_k - u), \varphi_k) + (T_\sigma u, \varphi_k - u)| \\
&\leq \|T_\sigma(\varphi_k - u)\|_{-m,2} \|\varphi_k\|_{m,2} + \|T_\sigma u\|_{-m,2} \|\varphi_k - u\|_{m,2} \\
&\leq C'\|\varphi_k - u\|_{m,2} \|\varphi_k\|_{m,2} + \|u\|_{m,2} \|\varphi_k - u\|_{m,2} \to 0
\end{aligned}
$$

as $k \to \infty$. Therefore

$$
\mathrm{Re}\,(T_\sigma u, u) = \lim_{k \to \infty} \mathrm{Re}\,(T_\sigma \varphi_k, \varphi_k) \geq C \lim_{k \to \infty} \|\varphi_k\|_{m,2}^2 = \|u\|_{m,2}^2.
$$

$\square$

We also need the following fact, which is a generalization of the Riesz representation theorem.

Theorem 18.4. (Lax–Milgram Lemma) *Let X be a complex and separable Hilbert space with inner product and norm denoted by $(\,,\,)_X$ and $\|\,\|_X$ respectively. Let B be a bilinear mapping on X such that there exist positive constants C_1 and C_2 for which*

$$
|B(x,y)| \leq C_1 \|x\|_X \|y\|_X, \quad x, y \in X, \tag{18.2}
$$

and

$$
|B(x,x)| \geq C_2 \|x\|_X^2, \quad x \in X. \tag{18.3}
$$

Then for every bounded linear functional f on X, there exists a unique vector $y \in X$ such that

$$
f(x) = B(x,y), \quad x \in X.
$$

Proof For a fixed vector y in X, we see from (18.2) that $B(\cdot, y)$ is a bounded linear functional on X. Using the Riesz representation theorem, there exists a unique vector $z(y) \in X$ such that

$$
B(x,y) = (x, z(y))_X, \quad x \in X.
$$

We observe that the mapping

$$
X \ni y \mapsto z(y) \in X
$$

is linear. Indeed, let y_1 and y_2 be vectors in X and let c_1 and c_2 be complex numbers. Since B is bilinear,

$$\begin{aligned}
(x, z(c_1 y_1 + c_2 y_2))_X &= B(x, c_1 y_1 + c_2 y_2) \\
&= \overline{c_1} B(x, y_1) + \overline{c_2} B(x, y_2) \\
&= \overline{c_1}(x, z(y_1))_X + \overline{c_2}(x, z(y_2))_X \\
&= (x, c_1 z(y_1))_X + (x, c_2 z(y_2))_X \\
&= (x, c_1 z(y_1)_X + c_2 z(y_2))_X, \quad x \in X.
\end{aligned}$$

Therefore

$$z(c_1 y_1 + c_2 y_2) = c_1 z(y_1) + c_2 z(y_2)$$

and linearity is established. Let M be the subspace of X given by

$$M = \{z(y) : y \in X\}.$$

Then M is a closed subspace of X. Indeed, let $\{z(y_k)\}$ be a sequence in M such that

$$z(y_k) \to z$$

as $k \to \infty$. Then for $j, k = 1, 2, \ldots$,

$$B(x, y_j - y_k) = (x, z(y_j) - z(y_k))_X, \quad x \in X.$$

By (18.3) and the Cauchy–Schwarz inequality, we get for $j, k = 1, 2, \ldots$,

$$\begin{aligned}
C_2 \|y_j - y_k\|_X^2 &\le |B(y_j - y_k, y_j - y_k)| \\
&= |(y_j - y_k, z(y_j) - z(y_k))_X| \\
&\le \|y_j - y_k\|_X \|z(y_j) - z(y_k)\|_X.
\end{aligned}$$

So, for $j, k = 1, 2, \ldots$,

$$C_2 \|y_j - y_k\|_X \le \|z(y_j) - z(y_k)\|_X.$$

Thus, $\{y_k\}$ is a Cauchy sequence in X and hence

$$y_k \to y$$

for some y in X as $k \to \infty$. By (18.2),

$$|B(x, y_k - y)| \le C_1 \|x\|_X \|y_k - y\|_X \to 0$$

as $k \to \infty$ for all $x \in X$, which gives

$$B(x, y_k) \to B(x, y)$$

as $k \to \infty$. Furthermore, for all $x \in X$,

$$(x, z(y_k))_X \to (x, z)_X$$

as $k \to \infty$. Since

$$B(x, y_k) = (x, z(y_k))_X, \quad x \in X,$$

for $k = 1, 2, \ldots$, it follows that

$$B(x, y) = (x, z)_X, \quad x \in X.$$

Therefore $z \in M$ and this proves that M is closed. Now, $M = X$. To see this, let us assume that M is a proper (and closed) subspace of X. Then there exists a nonzero vector $x \in X$ such that

$$B(x, y) = (x, z(y))_X = 0, \quad y \in X.$$

If we let $y = x$, then by (18.3), $x = 0$ and this contradiction implies that $M = X$. By the Riesz representation theorem, there exists a unique vector $w \in X$ such that

$$f(x) = (x, w)_X, \quad x \in X.$$

Since $X = M$, we can find a vector $y \in X$ such that $w = z(y)$. Therefore

$$f(x) = (x, w)_X = (x, z(y))_X = B(x, y), \quad x \in X.$$

The uniqueness of y follows again from (18.2). $\qquad\square$

Proof of Theorem 18.2 Let $B : H^{m,2} \times H^{m,2} \to \mathbb{C}$ be the bilinear mapping defined by

$$B(u, v) = (u, T_\sigma v), \quad u, v \in H^{m,2}.$$

Then for all u and v in $H^{m,2}$, we see by means of Exercise 12.14 that

$$|B(u, v)| \leq \|u\|_{m,2} \|T_\sigma v\|_{-m,2} \leq \|u\|_{m,2} \|v\|_{m,2}$$

and, by Lemma 18.3,

$$|B(u, u)| \geq |(T_\sigma u, u)| \geq C\|u\|_{m,2}^2, \quad u \in H^{m,2}.$$

Let $f \in L^2(\mathbb{R}^n)$. Then we define the linear functional $F : H^{m,2} \to \mathbb{C}$ by

$$F(w) = (w, f), \quad w \in H^{m,2}.$$

It is a bounded linear functional because, by Exercise 12.14 and Theorem 12.5,

$$|F(w)| = |(w, f)| \leq \|w\|_{m,2} \|f\|_{-m,2} \leq \|f\|_2 \|w\|_{m,2}, \quad w \in H^{m,2}.$$

So, by the Lax-Milgram lemma, we get a unique function u in $H^{m,2}$ such that

$$F(w) = B(w, u), \quad w \in H^{m,2},$$

or equivalently,

$$(w, f) = (w, T_\sigma u), \quad w \in H^{m,2}.$$

So, u is a weak solution in $L^2(\mathbb{R}^n)$. Since σ is elliptic, it follows from Exercise 18.2 that u is a strong solution in $L^2(\mathbb{R}^n)$. □

Another Proof of Theorem 18.2 By Theorem 12.5, the inequality (18.1) and the Cauchy–Schwarz inequality, we get for all functions $\varphi \in \mathcal{S}$,

$$\|\varphi\|_2^2 \leq \|\varphi\|_{m,2}^2 \leq \frac{1}{C} \|\varphi\|_2 \|T_\sigma^* \varphi\|_2$$

and hence

$$\|\varphi\|_2 \leq \frac{1}{C} \|T_\sigma^* \varphi\|_2.$$

Let $f \in L^2(\mathbb{R}^n)$. Then

$$|(f, \varphi)| \leq \|f\|_2 \|\varphi\|_2 \leq \frac{1}{C} \|f\|_2 \|T_\sigma^* \varphi\|_2, \quad \varphi \in \mathcal{S}.$$

So, by Lemma 18.3, the pseudo-differential equation $T_\sigma u = f$ on $\mathbb{R}^n$ has a weak solution u in $L^2(\mathbb{R}^n)$. Since σ is elliptic, it follows that u is also a strong solution in $L^2(\mathbb{R}^n)$. Let v be another strong solution in $L^2(\mathbb{R}^n)$. Then, by Theorem 18.3,

$$\|u - v\|_{m,2}^2 \leq \frac{1}{C} \operatorname{Re}(T_\sigma(u - v), u - v) = 0.$$

So, $u = v$ and u is the unique solution. □

We can now give sufficient conditions in terms of the symbols σ for the existence and uniqueness of strong solutions in $L^2(\mathbb{R}^n)$ for pseudo-differential operators T_σ.

Theorem 18.5. *Let $\sigma \in S^{2m}$, $m > 0$, be a strongly elliptic symbol. Then there exists a real number λ_0 such that for all f in $L^2(\mathbb{R}^n)$ and $\lambda \geq \lambda_0$, the pseudo-differential equation $(T_\sigma + \lambda I)u = f$ on $\mathbb{R}^n$, where I is the identity operator on $L^2(\mathbb{R}^n)$, has a unique strong solution u in $L^2(\mathbb{R}^n)$.*

Proof By Gårding's inequality, there exist constants A and λ_0 such that $A > 0$ and

$$\operatorname{Re}(T_\sigma \varphi, \varphi) \geq A\|\varphi\|_{m,2}^2 - \lambda_0 \|\varphi\|_2^2, \quad \varphi \in \mathcal{S}.$$

Then for $\lambda \geq \lambda_0$,

$$\text{Re}\left((T_\sigma + \lambda I)\varphi, \varphi\right) \geq A\|\varphi\|_{m,2}^2 + (\lambda - \lambda_0)\|\varphi\|_2^2 \geq A\|\varphi\|_{m,2}^2, \quad \varphi \in \mathcal{S}.$$

Thus, by Theorem 18.2, the proof is complete. $\qquad\square$

The following theorem is a result on the existence and uniqueness of strong solutions in $L^p(\mathbb{R}^n)$.

Theorem 18.6. *Let $\sigma \in S^m$, $m > 0$, be an elliptic symbol such that σ is independent of x in $\mathbb{R}^n$ and*

$$\sigma(\xi) \neq 0, \quad \xi \in \mathbb{R}^n.$$

Then for every function $f \in L^p(\mathbb{R}^n)$, $1 < p < \infty$, the pseudo-differential equation $T_\sigma u = f$ on $\mathbb{R}^n$ has a unique strong solution u in $L^p(\mathbb{R}^n)$.

Proof By Exercise 18.3, it is sufficient to prove that there exists a positive constant C such that

$$\|\varphi\|_{p'} \leq C\|T_{\overline{\sigma}}\varphi\|_{p'}, \quad \varphi \in \mathcal{S}.$$

Let $\tau = 1/\overline{\sigma}$. Then for all multi-indices α, we use (1.4) to obtain

$$\partial^\alpha \tau = \sum C_{\alpha^{(1)},\ldots,\alpha^{(k)}} \frac{(\partial^{\alpha^{(1)}}\overline{\sigma})\cdots(\partial^{\alpha^{(k)}}\overline{\sigma})}{\overline{\sigma}^{k+1}},$$

where $C_{\alpha^{(1)},\ldots,\alpha^{(k)}}$ is a constant depending on $\alpha^{(1)},\ldots,\alpha^{(k)}$ and the sum is taken over all multi-indices $\alpha^{(1)},\ldots,\alpha^{(k)}$ that partition α. Thus, we can find positive constants $C_{\alpha^{(1)}},\ldots,C_{\alpha^{(k)}}$ such that

$$|(\partial^\alpha \tau)(\xi)| \leq \sum |C_{\alpha^{(1)},\ldots,\alpha^{(k)}}| \frac{C_{\alpha^{(1)}}\cdots C_{\alpha^{(k)}}(1+|\xi|)^{km-|\alpha|}}{|\sigma(\xi)|^{k+1}}$$

for all $\xi \in \mathbb{R}^n$. Since σ is elliptic, we can find positive constants C and R such that

$$|\sigma(\xi)| \geq C(1+|\xi|)^m, \quad |\xi| \geq R.$$

Since $\frac{|\sigma|}{(1+|\ |)^m}$ is a continuous and positive function on the compact set $\{\xi \in \mathbb{R}^n : |\xi| \leq R\}$, it follows that there exists a positive number δ such that

$$|\sigma(\xi)| \geq \delta(1+|\xi|)^m, \quad |\xi| \leq R.$$

Therefore there exists a positive constant C' such that

$$|\sigma(\xi)| \geq C'(1+|\xi|)^m, \quad \xi \in \mathbb{R}^n.$$

So, there exists a positive constant C'' such that

$$|(\partial^\alpha \tau)(\xi)| \le C''(1 + |\xi|)^{-m-|\alpha|}, \quad \xi \in \mathbb{R}^n.$$

Therefore $\tau \in S^{-m}$. By Theorem 12.9 and the Sobolev embedding theorem in Chapter 12, we get a positive constant C'''' such that

$$\|\varphi\|_{p'} = \|T_\tau T_{\bar{\sigma}}\varphi\|_{p'} \le C'''\|T_{\bar{\sigma}}\varphi\|_{p'}, \quad \varphi \in \mathcal{S}.$$

It remains to prove uniqueness. Let u and v be strong solutions in $L^p(\mathbb{R}^n)$. Let $w = u - v$. Then $w \in H^{m,p}$ and $T_\sigma w = 0$ on $\mathbb{R}^n$. Thus,

$$\sigma \hat{w} = 0$$

in the sense of distributions. So,

$$\hat{w}(\sigma\varphi) = (\sigma\hat{w})(\varphi) = 0, \quad \varphi \in \mathcal{S}. \tag{18.4}$$

For all $\psi \in \mathcal{S}$, we can find a function $\varphi \in \mathcal{S}$ such that $\sigma\varphi = \psi$. That this can be done is Exercise 18.5. Thus, by (18.4),

$$\hat{w}(\psi) = 0, \quad \psi \in \mathcal{S}.$$

So, $\hat{w} = 0$ and hence $w = 0$. Therefore $u = v$ and uniqueness is proved. $\square$

Exercises

18.1. Let $\sigma \in S^m$, $m > 0$. Let $f \in L^p(\mathbb{R}^n)$ for $1 < p < \infty$. Prove that a strong solution u in $L^p(\mathbb{R}^n)$ of the equation $T_\sigma u = f$ on $\mathbb{R}^n$ is also a weak solution in $L^p(\mathbb{R}^n)$.

18.2. Let $\sigma \in S^m$, $m > 0$, be an elliptic symbol. Let $f \in L^p(\mathbb{R}^n)$, where $1 < p < \infty$. Prove that every weak solution u in $L^p(\mathbb{R}^n)$ of the equation $T_\sigma u = f$ on $\mathbb{R}^n$ is also a strong solution in $L^p(\mathbb{R}^n)$.

18.3. Let $\sigma \in S^m$, $m > 0$, be elliptic and such that there exists a positive constant C for which

$$\|\varphi\|_{p'} \le C\|T_\sigma^*\varphi\|_{p'}, \quad \varphi \in \mathcal{S}.$$

Prove that the pseudo-differential equation $T_\sigma u = f$ on $\mathbb{R}^n$ has a strong solution u in $L^p(\mathbb{R}^n)$ for every function $f \in L^p(\mathbb{R}^n)$, $1 < p < \infty$.

18.4. Let σ be as in Exercise 18.3 and such that $\{T_\sigma^*\varphi : \varphi \in \mathcal{S}\}$ is dense in $L^{p'}(\mathbb{R}^n)$. Prove that the equation $T_\sigma u = f$ on $\mathbb{R}^n$ has a unqiue strong solution u in $L^p(\mathbb{R}^n)$ for every function $f \in L^p(\mathbb{R}^n)$, $1 < p < \infty$.

18.5. Let $\sigma \in S^m$, $m > 0$, be an elliptic symbol such that σ is independent of x in $\mathbb{R}^n$ and

$$\sigma(\xi) \ne 0, \quad \xi \in \mathbb{R}^n.$$

Prove that for all functions $\psi \in \mathcal{S}$, there exists a function $\varphi \in \mathcal{S}$ such that $\sigma\varphi = \psi$.

Chapter 19

One-Parameter Semigroups Generated by Pseudo-Differential Operators

Let A be a closed linear operator from a complex Banach space X into X with dense domain $\mathcal{D}(A)$. Let $f \in X$. Then we are interested in the initial value problem for the heat equation governed by A given by

$$\begin{cases} u'(t) = A(u(t)), & t > 0, \\ u(0) = f, \end{cases} \tag{19.1}$$

where $u : [0, \infty) \to X$ and

$$u'(t) = \lim_{h \to 0} \frac{u(t+h) - u(t)}{h}$$

if the limit in X exists. The questions to be answered are related to the existence and uniqueness of a global solution $u : [0, \infty) \to X$.

It is intuitively clear that the solution $u : [0, \infty) \to X$ is given *formally* by

$$u(t) = e^{tA} f, \quad t \geq 0.$$

What is e^{tA} when A is a closed linear operator densely defined on a complex Banach space X? The aim of this chapter is to answer this question.

A family $\{T(t) : t \geq 0\}$ of bounded linear operators on X is said to be a *one-parameter semigroup* if
(i) $T(0) = I$, where I is the identity operator on X,
(ii) $T(s)T(t) = T(s + t), \quad s, t \geq 0$,
(iii) $T(t)x \to x$ in X as $t \to 0+$ for every x in X.

Let $\{T(t) : t \geq 0\}$ be a one-parameter semigroup on X. We denote by $\mathcal{D}(A)$ the set of all elements x in X such that $\lim_{t \to 0+} \frac{T(t)x - x}{t}$ exists in X.

Proposition 19.1. $\mathcal{D}(A)$ *is a dense subspace of* X.

Proof Let W be the subspace of X given by

$$W = \left\{ x_s = \frac{1}{s} \int_0^s T(\delta) x \, d\delta : x \in X, s > 0 \right\}.$$

Then W is dense in X simply because $x_s \to x$ in X for all $x \in X$ as $s \to 0+$. For $t > 0$, let

$$A_t = \frac{T(t) - I}{t}.$$

Let s and t be positive numbers. Then for $t < s$,

$$
\begin{aligned}
A_t x_s &= \frac{T(t) \int_0^s T(\delta) x \, d\delta - \int_0^s T(\delta) x \, d\delta}{st} \\
&= \frac{\int_0^s T(t + \delta) x \, d\delta - \int_0^s T(\delta) x \, d\delta}{st} \\
&= \frac{\int_t^{s+t} T(\delta) x \, d\delta - \int_0^s T(\delta) x \, d\delta}{st} \\
&= \frac{\int_s^{s+t} T(\delta) x \, d\delta - \int_0^t T(\delta) x \, d\delta}{st} \\
&= A_s x_t.
\end{aligned}
$$

Thus, $A_t x_s \to A_s x$ in X as $t \to 0+$. Therefore $W \subseteq \mathcal{D}(A)$ and hence $\mathcal{D}(A)$ is dense in X. □

The linear operator A from X into X with domain $\mathcal{D}(A)$ defined by

$$Ax = \lim_{t \to 0+} \frac{T(t)x - x}{t}, \quad x \in \mathcal{D}(A),$$

where the limit is again understood to take place in X, is called the *infinitesimal generator* of the one-parameter semigroup $\{T(t) : t \geq 0\}$.

Proposition 19.2. *The infinitesimal generator A of a one-parameter semigroup $\{T(t) : t \geq 0\}$ on a complex Banach space X is a closed linear operator.*

We need two lemmas to prove Proposition 19.2.

Lemma 19.3. *Let A be the infinitesimal generator of a one-parameter semigroup $\{T(t) : t \geq 0\}$ of bounded linear operators on a complex Banach space X. Then for all $t \in [0, \infty)$ and $x \in \mathcal{D}(A)$, $T(t)x \in \mathcal{D}(A)$ and*

$$AT(t)x = T(t)Ax, \quad x \in \mathcal{D}(A).$$

Proof Let $x \in \mathcal{D}(A)$. Then for all $t \in [0, \infty)$,

$$T(t)A_h x = A_h T(t)x, \quad h > 0.$$

Therefore

$$\lim_{h \to 0} A_h T(t)x = \lim_{h \to 0} T(t)A_h x = T(t) \lim_{h \to 0} A_h x = T(t)Ax.$$

Thus, $T(t)x \in \mathcal{D}(A)$ and

$$AT(t)x = T(t)Ax,$$

as asserted. $\qquad\qquad\qquad\qquad\qquad\qquad\qquad\qquad\qquad\qquad\qquad$ $\square$

Lemma 19.4. *Let $x \in \mathcal{D}(A)$, where A is the infinitesimal generator of a one-parameter semigroup $\{T(t) : t \geq 0\}$ on a complex Banach space X. Then for all $t > 0$,*

$$T(t)x - x = \int_0^t T(s)Ax \, ds.$$

Proof Let f be a bounded linear functional on X. Then we define the function $F : [0, \infty) \to \mathbb{C}$ by

$$F(t) = f \left(T(t)x - x - \int_0^t T(s)Ax \, ds \right), \quad t \geq 0.$$

So, for $t \geq 0$, the right-hand derivative $F'(t+)$ of F at t is given by

$$F'(t+) = f\left(AT(t)x - T(t)Ax\right)$$

in view of the definition of the infinitesimal generator and the fundamental theorem of calculus. By Lemma 19.3,

$$F'(t+) = 0, \quad t \in [0, \infty).$$

Thus, $F(t)$ is a constant for all $t \in [0, \infty)$. Since $F(0) = 0$, it follows that

$$F(t) = 0, \quad t \geq 0.$$

Thus, using the Hahn–Banach theorem, we see that

$$T(t)x - x = \int_0^t T(s)Ax \, ds, \quad t \geq 0.$$

$\qquad\qquad\qquad\qquad\qquad\qquad\qquad\qquad\qquad\qquad\qquad\qquad$ $\square$

Proof of Proposition 19.2 Let $\{x_j\}$ be a sequence in $\mathcal{D}(A)$ such that

$$x_j \to x$$

and

$$Ax_j \to y$$

in X as $j \to \infty$. Then, by Lemma 19.4,

$$T(t)x - x = \lim_{j \to \infty} (T(t)x_j - x_j) = \lim_{j \to \infty} \int_0^t T(s)Ax_j \, ds = \int_0^t T_s y \, ds.$$

Hence

$$\lim_{t \to 0+} \frac{T(t)x - x}{t} = \lim_{t \to 0+} \int_0^t T_s y \, ds = y.$$

Therefore $x \in \mathcal{D}(A)$ and $Ax = y$. Thus, A is a closed linear operator. $\square$

When is a closed linear operator A from a complex Banach space X into X with dense domain $\mathcal{D}(A)$ the infinitesimal generator of a one-parameter semigroup on X? The answer is given as the content of the celebrated Hille–Yosida–Phillips theorem.

We first introduce some notation and terminology. Let A be a closed linear operator from a complex Banach space X into X with dense domain $\mathcal{D}(A)$. We define the *resolvent set* $\rho(A)$ of A to be the set of all complex numbers λ such that $\lambda I - A : \mathcal{D}(A) \to X$ is bijective, where I is the identity operator on X. If $\lambda \in \rho(A)$, then the bounded linear operator $(\lambda I - A)^{-1} : X \to X$ is called the *resolvent* of A at λ and is often denoted by $R(\lambda; A)$. The norm a bounded linear operator A from a complex Banach space X into another complex Banach space Y is denoted by $\|A\|$.

Theorem 19.5. (Hille–Yosida–Phillips Theorem) *Let A be a closed linear operator from a complex Banach space X into X with dense domain $\mathcal{D}(A)$. Then A is the infinitesimal generator of a one-parameter semigroup $\{T(t) : t \geq 0\}$ on X if and only if we can find a positive number M and a real number ω such that*

$$\{\lambda \in \mathbb{R} : \lambda > \omega\} \subseteq \rho(A)$$

and

$$\|R(\lambda; A)^n\| \leq \frac{M}{(\lambda - \omega)^n}$$

for all λ in (ω, ∞) and all positive integers n.

Proof Let us first prove the sufficiency. For $\lambda > \omega$, we define the bounded linear operator $B_\lambda : X \to X$ by

$$B_\lambda = -\lambda(I - \lambda R(\lambda; A)).$$

Then for all $t \in \mathbb{R}$,

$$e^{tB_\lambda} = e^{-\lambda t} \sum_{n=0}^{\infty} \frac{(\lambda^2 t)^n}{n!} R(\lambda; A)^n$$

and therefore

$$\|e^{tB_\lambda}\| \le M e^{-\lambda t} \sum_{n=0}^{\infty} \frac{(\lambda^2 t)^n}{n!(\lambda - \omega)^n} = M e^{\frac{t\omega\lambda}{\lambda - \omega}}.$$

So, for every ω_1 in (ω, ∞),

$$\|e^{tB_\lambda}\| < M e^{t\omega_1} \tag{19.2}$$

for all sufficiently large λ. Now, we claim that

$$\lim_{\lambda \to \infty} B_\lambda x = Ax, \quad x \in \mathcal{D}(A). \tag{19.3}$$

Indeed, let $x \in \mathcal{D}(A)$. Then

$$\begin{aligned}
\lambda R(\lambda; A)x - x &= \lambda R(\lambda; A)x - R(\lambda; A)Ax + R(\lambda; A)Ax - x \\
&= R(\lambda; A)(\lambda I - A)x + R(\lambda; A)Ax - x \\
&= x + R(\lambda; A)Ax - x \\
&= R(\lambda; A)Ax
\end{aligned}$$

and hence

$$\|\lambda R(\lambda; A)x - x\|_X = \|R(\lambda; A)Ax\|_X \le \frac{M}{\lambda - \omega}\|Ax\|_X \to 0$$

as $\lambda \to \infty$. Moreover, for all sufficiently large λ,

$$\|\lambda R(\lambda; A)\| \le M \frac{\lambda}{\lambda - \omega} < 2M.$$

Since $\mathcal{D}(A)$ is dense in X, it follows from a standard limiting argument that for every x in X,

$$\lambda R(\lambda; A)x \to x$$

in X as $\lambda \to \infty$. But for all x in $\mathcal{D}(A)$,

$$\begin{aligned}
B_\lambda x &= -\lambda x + \lambda^2 R(\lambda; A)x \\
&= -\lambda x + \lambda R(\lambda; A)\lambda x - \lambda R(\lambda; A)Ax + \lambda R(\lambda; A)Ax \\
&= \lambda R(\lambda; A)Ax
\end{aligned}$$

and hence we get

$$B_\lambda x = \lambda R(\lambda; A)Ax \to Ax$$

as $\lambda \to \infty$. For $\lambda > \omega$, let

$$S_\lambda(t) = e^{tB_\lambda}, \quad t \in \mathbb{R}.$$

Using the first resolvent formula in Exercise 19.4, we know that

$$R(\lambda; A)R(\mu; A) = R(\mu; A)R(\lambda; A), \quad \lambda, \mu > \omega.$$

Thus, $B_\lambda B_\mu = B_\mu B_\lambda$ and we get

$$B_\mu S_\lambda(t) = S_\lambda(t)B_\mu$$

for all λ and μ in (ω, ∞), and all t in $\mathbb{R}$. Hence for all $x \in \mathcal{D}(A)$ and $t \in [0, \infty)$,

$$
\begin{aligned}
S_\lambda(t)x - S_\mu(t)x &= \int_0^t \frac{d}{ds}[S_\mu(t-s)S_\lambda(s)x]\, ds \\
&= \int_0^t S_\mu(t-s)S_\lambda(s)(B_\lambda - B_\mu)x\, ds. \quad (19.4)
\end{aligned}
$$

By (19.2) and (19.4), there exists a positive constant λ_0 such that

$$\|S_\lambda(t)x - S_\mu(t)x\|_X \le M^2 t e^{t\omega_1}\|B_\lambda x - B_\mu x\|_X, \quad t \ge 0, \quad (19.5)$$

for all λ and μ in (λ_0, ∞). By (19.3) and (19.5), we see that for all x in $\mathcal{D}(A)$ and all t in $[0, \infty)$,

$$\|S_\lambda(t)x - S_\mu(t)x\|_X \to 0$$

as λ and μ tend to ∞. So, in view of (19.2), we see that for all x in X and all t in $[0, \infty)$,

$$\|S_\lambda(t)x - S_\mu(x)\|_X \to 0$$

as λ and μ tend to ∞. Therefore for all x in X and all t in $[0, \infty)$, we define $T(t)x$ by

$$T(t)x = \lim_{\lambda \to \infty} S_\lambda(t)x.$$

It is then easy to see that $T(t)$ is a bounded linear operator on X and

$$\|T(t)\| \le Me^{t\omega_1} \quad (19.6)$$

for all t in $[0, \infty)$. Let $x \in \mathcal{D}(A)$. Then, by (19.5), $S_\lambda(t)x \to T(t)x$ uniformly with respect to t on compact subsets of $[0, \infty)$ as $\lambda \to \infty$. Therefore for all x in X, the mapping

$$[0, \infty) \ni t \mapsto T(t)x \in X$$

is continuous. To check the semigroup property, we first note that

$$T(0) = \lim_{\lambda \to \infty} S_\lambda(0) = I.$$

Moreover, for $0 \leq s, t < \infty$ and all x in X,

$$T(s+t)x = \lim_{\lambda \to \infty} S_\lambda(s+t)x = \lim_{\lambda \to \infty} S_\lambda(s)S_\lambda(t)x$$

and, by (19.2),

$$\|S_\lambda(s)S_\lambda(t)x - T(s)T(t)x\|_X$$
$$= \|S_\lambda(s)S_\lambda(t)x - S_\lambda(s)T(t)x + S_\lambda(s)T(t)x - T(s)T(t)x\|_X$$
$$\leq \|S_\lambda(s)\| \, \|S_\lambda(t)x - T(t)x\|_X + \|(S_\lambda(s) - T(s))T(t)x\|_X$$
$$\leq Me^{s\omega_1}\|S_\lambda(t)x - T(t)x\|_X + \|(S_\lambda(s) - T(s))T(t)x\|_X \to 0$$

as $\lambda \to \infty$. It remains to prove that A is equal to the infinitesimal generator B of the one-parameter semigroup $\{T(t) : t \geq 0\}$. Let $x \in X$. Then

$$S_\lambda(t)x - x = \int_0^t \frac{d}{ds}S_\lambda(s)x \, ds = \int_0^t S_\lambda(s)B_\lambda x \, ds. \qquad (19.7)$$

For all x in $\mathcal{D}(A)$ and all s in $[0, \infty)$,

$$\|S_\lambda(s)B_\lambda x - T(s)Ax\|_X$$
$$= \|S_\lambda(s)B_\lambda x - S_\lambda(s)Ax + S_\lambda(s)Ax - T(s)Ax\|_X$$
$$\leq \|S_\lambda(s)\| \, \|B_\lambda x - Ax\|_X + \|(S_\lambda(s) - T(s))Ax\|_X \to 0$$

uniformly with respect to s on compact subsets of $[0, \infty)$ as $\lambda \to \infty$. Thus, by (19.7),

$$T(t)x - x = \int_0^t T(s)Ax \, ds, \qquad x \in \mathcal{D}(A).$$

Therefore

$$Bx = \lim_{t \to 0+} \frac{T(t)x - x}{t} = \lim_{t \to 0+} \int_0^t T(s)Ax \, ds = Ax, \qquad x \in \mathcal{D}(A).$$

Hence B is an extension of A. So, we only need to prove that $\mathcal{D}(A) = \mathcal{D}(B)$. To this end, we claim that there exists a real number λ_1 such that

$$\lambda \in \rho(B), \qquad \lambda > \lambda_1. \qquad (19.8)$$

Assuming the claim for a moment, then there exists a real number λ such that

$$\lambda \in \rho(A) \cap \rho(B).$$

Thus,

$$(\lambda I - B)\mathcal{D}(A) = (\lambda I - A)\mathcal{D}(A) = X$$

and

$$(\lambda I - B)\mathcal{D}(B) = X.$$

So, if $x \in \mathcal{D}(B)$, then there exists an element z in $\mathcal{D}(A)$ such that

$$(\lambda I - B)z = (\lambda I - B)x.$$

Since $\lambda I - B$ is one-to-one, $z = x \in \mathcal{D}(A)$. It remains to prove the claim, i.e., (19.8). For $\lambda > \omega_1$, we define $R(\lambda)x$ for every x in X by

$$R(\lambda)x = \int_0^\infty e^{-\lambda t} T(t)x \, dt.$$

That the integral exists is a consequence of (19.6). It also follows from (19.6) that $R(\lambda)$ is a bounded linear operator on X. Now, for $\lambda > \omega_1$ and all x in X,

$$
\begin{aligned}
& B_h R(\lambda)x \\
&= \frac{T(h)R(\lambda)x - R(\lambda)x}{h} \\
&= \frac{1}{h}\int_0^\infty e^{-\lambda t} T(t+h)x \, dt - \frac{1}{h}\int_0^\infty e^{-\lambda t} T(t)x \, dt \\
&= \frac{1}{h}\int_h^\infty e^{-\lambda(s-h)} T(s)x \, ds - \frac{1}{h}\int_0^\infty e^{-\lambda t} T(t)x \, dt \\
&= \frac{e^{\lambda h}}{h}\int_0^\infty e^{-\lambda s} T(s)x \, ds - \frac{1}{h}\int_0^\infty e^{-\lambda t} T(t)x \, dt - \frac{e^{\lambda h}}{h}\int_0^h e^{-\lambda t} T(t)x \, dt \\
&= \frac{e^{\lambda h} - 1}{h}\int_0^\infty e^{-\lambda t} T(t)x \, dt - \frac{e^{\lambda h}}{h}\int_0^h e^{-\lambda t} T(t)x \, dt \\
&\to \lambda R(\lambda)x - x
\end{aligned}
$$

as $h \to 0+$. Thus, for $\lambda > \omega_1$ and all x in X.

$$R(\lambda)x \in \mathcal{D}(B)$$

and

$$BR(\lambda)x = \lambda R(\lambda)x - x,$$

which is the same as

$$(\lambda I - B)R(\lambda)x = x. \tag{19.9}$$

So, if $\lambda > \omega_1$ and $x \in \mathcal{D}(B)$, then

$$R(\lambda)x \in \mathcal{D}(B)$$

and

$$BR(\lambda)x = B \int_0^\infty e^{-\lambda t} T(t)x \, dt$$

$$= \int_0^\infty e^{-\lambda t} BT(t)x \, dt$$

$$= \int_0^\infty e^{-\lambda t} T(t)Bx \, dt$$

$$= R(\lambda)Bx. \tag{19.10}$$

By (19.9) and (19.10),

$$R(\lambda)(\lambda I - B)x = x.$$

So, for $\lambda > \omega_1$, $(\lambda I - B)^{-1} = R(\lambda)$ and the proof of the sufficiency part of the theorem is complete. As for the necessity, we let $g : [0, \infty) \to \mathbb{R}$ be the function defined by

$$g(t) = \ln \|T(t)\|, \quad t \in [0, \infty).$$

Then g is subadditive. Indeed, for all s and t in $[0, \infty)$,

$$g(s + t) = \ln \|T(s + t)\| = \ln (\|T(s)T(t)\|)$$
$$\leq \ln (\|T(s)\| \|T(t)\|) = \ln \|T(s)\| + \ln \|T(t)\| = g(s) + g(t).$$

Let $t_0 > 0$. Then for all t in $[0, \infty)$, we write $t = nt_0 + s$, where n is an integer depending on t and $0 \leq s < t_0$. By the subadditivity of g, we get

$$\frac{g(t)}{t} \leq \frac{ng(t_0)}{t} + \frac{g(s)}{t} \to \frac{g(t_0)}{t_0}$$

as $t \to \infty$. So,

$$\overline{\lim}_{t \to \infty} \frac{g(t)}{t} \leq \frac{g(t_0)}{t_0}, \quad t_0 > 0,$$

and hence

$$\overline{\lim}_{t \to \infty} \frac{g(t)}{t} \leq \inf_{t_0 > 0} \frac{g(t_0)}{t_0}.$$

Let $\delta > \inf_{t_0 > 0} \frac{g(t_0)}{t_0}$. Then there exists a positive number R such that

$$\sup_{t > R} \frac{g(t)}{t} < \delta.$$

Thus,

$$\ln \|T(t)\| \leq \delta t, \quad t > R,$$

or

$$\|T(t)\| < e^{\delta t}, \quad t > R.$$

Since $\frac{\|T(t)\|}{e^{\delta t}}$ is a continuous function of t on the interval $[0, R]$, there exists a positive constant C for which

$$\|T(t)\| \le Ce^{\delta t}, \quad t \in [0, R].$$

Hence there is a positive constant M such that

$$\|T(t)\| \le Me^{\delta t}, \quad t \in [0, \infty). \tag{19.11}$$

Now, for $\lambda > \delta$ and for all x in X, we define $R(\lambda)x$ by

$$R(\lambda)x = \int_0^\infty e^{-\lambda t} T(t)x \, dt.$$

Since A is the infinitesimal generator of the one-parameter semigroup $\{T(t) : t \ge 0\}$, we can repeat the analysis given above with A for B and δ for ω_1 to conclude that $(\delta, \infty) \subseteq \rho(A)$ and

$$R(\lambda; A) = \int_0^\infty e^{-\lambda t} T(t)x \, dt \tag{19.12}$$

for $\lambda > \delta$ and for all x in X. For all λ and μ in (δ, ∞), the first resolvent formula in Exercise 19.4 gives

$$R(\lambda; A) - R(\mu; A) = (\mu - \lambda)R(\lambda; A)R(\mu; A).$$

Thus,

$$\frac{d}{d\lambda} R(\lambda; A) = -R(\lambda; A)^2$$

and, by induction, we get

$$\frac{d^n}{d\lambda^n} R(\lambda; A) = (-1)^n n! R(\lambda; A)^{n+1}$$

for all λ in (δ, ∞). Differentiating both sides of (19.12) $n - 1$ times with respect to λ and using the preceding formula for $\frac{d^n}{d\lambda^n} R(\lambda; A)$, we get

$$R(\lambda; A)^n = \frac{1}{(n-1)!} \int_0^\infty e^{-\lambda t} t^{n-1} T(t)x \, dt \tag{19.13}$$

for $\lambda > \delta$ and for all x in X. So, by (19.11) and (19.13),

$$\|R(\lambda; A)^n\| \le \frac{M}{(n-1)!} \int_0^\infty t^{n-1} e^{-(\lambda-\delta)t} dt = \frac{M}{(\lambda-\delta)^n}$$

for all positive integers n. $\qquad\qquad\qquad\qquad\qquad\qquad\qquad\qquad\square$

Remark 19.6. We leave it as Exercise 19.6 to prove that the function $u : [0, \infty) \to X$ defined by

$$u(t) = T(t)f, \quad t \ge 0,$$

is the unique solution of the initial value problem (19.1). Thus, e^{tA} is the natural notation for $T(t)$ for $t \ge 0$.

As an application to pseudo-differential operators, we give the following theorem.

Theorem 19.7. *Let $\sigma \in S^{2m}$, $m > 0$, be a symbol such that we can find a positive constant C and a constant λ_0 for which*

$$\text{Re}\,(-T_\sigma\varphi, \varphi) \geq C\|\varphi\|_{m,2}^2 - \lambda_0\|\varphi\|_2^2, \quad \varphi \in \mathcal{S}. \tag{19.14}$$

Then the operator $T_{\sigma,0}$ is the infinitesimal generator of a one-parameter semigroup of bounded linear operators on $L^2(\mathbb{R}^n)$.

For a proof of Theorem 19.7, we need the following lemma.

Lemma 19.8. *Let $\lambda > \lambda_0$. Then for every $f \in L^2(\mathbb{R}^n)$, there exists a unique solution $u \in H^{2m,2}$ of the equation $(\lambda I - T_{\sigma,0})u = f$, where I is the identity operator on $L^2(\mathbb{R}^n)$. Moreover,*

$$\|(\lambda I - T_{\sigma,0})u\|_2 \geq (\lambda - \lambda_0)\|u\|_2, \quad u \in H^{2m,2}. \tag{19.15}$$

Proof Let $\lambda > \lambda_0$. Then, by (19.14),

$$\begin{aligned}
\text{Re}\,((\lambda I - T_\sigma)\varphi, \varphi) &= \text{Re}\,((\lambda_0 I - T_\sigma)\varphi, \varphi) + (\lambda - \lambda_0)\|\varphi\|_2^2 \\
&\geq C\|\varphi\|_{m,2}^2 + (\lambda - \lambda_0)\|\varphi\|_2^2 \\
&\geq (\lambda - \lambda_0)\|\varphi\|_2^2, \quad \varphi \in \mathcal{S}.
\end{aligned}$$

So, by a limiting argument,

$$\text{Re}\,((\lambda I - T_{\sigma,0})u, u) \geq (\lambda - \lambda_0)\|u\|_2^2, \quad u \in H^{2m,2}.$$

Thus, for every $f \in L^2(\mathbb{R}^n)$, we can use Theorem 18.2 to obtain a unique solution $u \in H^{2m,2}$ of the equation

$$(\lambda I - T_{\sigma,0})u = f.$$

Moreover,

$$\begin{aligned}
&\|(\lambda I - T_{\sigma,0})\varphi\|_2^2 \\
&= (\lambda - \lambda_0)^2\|\varphi\|_2^2 + 2(\lambda - \lambda_0)\text{Re}\,((\lambda_0 I - T_{\sigma,0})\varphi, \varphi) + \|\lambda_0 I - T_{\sigma,0}\varphi\|_2^2 \\
&\geq (\lambda - \lambda_0)^2\|\varphi\|_2^2, \quad \varphi \in \mathcal{S}.
\end{aligned}$$

Thus, by a standard limiting argument again,

$$\|(\lambda I - T_{\sigma,0})u\|_2 \geq (\lambda - \lambda_0)\|u\|_2, \quad u \in H^{2m,2}.$$

$\square$

Proof of Theorem 19.7 $T_{\sigma,0}$ is a closed and densely defined linear operator from $L^2(\mathbb{R}^n)$ into $L^2(\mathbb{R}^n)$. By Lemma 19.8, $(\lambda I - T_{\sigma,0})^{-1}$ exists for $\lambda > \lambda_0$. Let $\lambda > \lambda_0$. Then, by (19.15),

$$\|(\lambda I - T_{\lambda,0})^{-1}\| \leq (\lambda - \lambda_0)^{-1}.$$

Thus,

$$\|(\lambda I - T_{\sigma,0})^{-n}\| \leq \|(\lambda I - T_{\lambda,0})^{-1}\|^n \leq (\lambda - \lambda_0)^{-n}, \quad n = 1, 2, \ldots.$$

Hence, using the Hille–Yosida–Phillips theorem, the proof is complete. □

The following theorem gives a familiar class of symbols σ for which $-T_\sigma$ is the infinitesimal generator of a one-parameter semigroup of bounded linear operators on $L^2(\mathbb{R}^n)$.

Theorem 19.9. *Let $\sigma \in S^{2m}, m > 0$, be a strongly elliptic symbol. Then $-T_{\sigma,0}$ is the infinitesimal generator of a one-parameter semigroup of bounded linear operators on $L^2(\mathbb{R}^n)$.*

Proof By Gårding's inequality, i.e., Theorem 17.1, we can find a positive constant C and a constant C_s for every real number $s \geq \frac{1}{2}$ such that

$$\mathrm{Re}\,(T_\sigma\varphi, \varphi) \geq C\|\varphi\|_{m,2}^2 - C_s\|\varphi\|_{m-s,2}^2, \quad \varphi \in \mathcal{S}. \tag{19.16}$$

If $C_s \leq 0$, then

$$\mathrm{Re}\,(T_\sigma\varphi, \varphi) \geq C\|\varphi\|_{m,2}^2, \quad \varphi \in \mathcal{S}.$$

Now, suppose that $C_s > 0$. Let $\varepsilon \in (0, C/C_s)$. Since $m - s < m$, it follows from Erhling's inequality in Exercise 12.6 that there exists a positive constant C_ε for which

$$\|\varphi\|_{m-s,2}^2 \leq \varepsilon\|\varphi\|_{m,2}^2 + C_\varepsilon\|\varphi\|_2^2, \quad \varphi \in \mathcal{S}. \tag{19.17}$$

Thus, by (19.16) and (19.17),

$$\mathrm{Re}\,(T_\sigma\varphi, \varphi) \geq (C - C_s\varepsilon)\|\varphi\|_{m,2}^2 - C_s C_\varepsilon\|\varphi\|_2^2, \quad \varphi \in \mathcal{S}.$$

So, by Theorem 19.7, the operator $T_{\sigma,0}$ is the infinitesimal generator of a one-parameter semigroup of bounded linear operators on $L^2(\mathbb{R}^n)$. □

Exercises

19.1. Let A be a bounded linear operator from a complex Banach space X into X.
(i) Prove that for $t \geq 0$, the series $\sum_{k=0}^\infty \frac{t^k A^k}{k!}$ converges absolutely. (We denote the sum by e^{tA}.)
(ii) Prove that for all $x \in X$,

$$\lim_{h \to 0+} \frac{e^{(t+h)A}x - e^{tA}x}{h} = Ae^{tA}x, \quad t \geq 0.$$

(iii) Prove that for all $s, t \in [0, \infty)$,

$$e^{sA} e^{tA} = e^{(s+t)A}.$$

(iv) Prove that $e^{0A} = I$, where I is the identity operator on X.

(v) Prove that for all $t \in [0, \infty)$, $A e^{tA} = e^{tA} A$.

19.2. Let $\{T(t) : t \geq 0\}$ be a one-parameter semigroup on a complex Banach space X. Prove that the subspace W of X defined by

$$W = \left\{ x_s = \frac{1}{s} \int_0^s T(\delta) x \, d\delta : x \in X, \, s > 0 \right\}$$

is dense in X.

19.3. Let A be a closed linear operator from a complex Banach space X into X with dense domain. Prove that $\rho(A)$ is an open subset of the complex plane $\mathbb{C}$.

19.4. Let A be a closed linear operator from a complex Banach space X into X with dense domain. Prove that for all λ_1 and λ_2 in $\rho(A)$,

$$R(\lambda_1; A) - R(\lambda_2; A) = (\lambda_2 - \lambda_1) R(\lambda_1; A) R(\lambda_2; A).$$

(This is known as the *first resolvent formula*.)

19.5. Prove (19.4) in the proof of the Hille–Yosida–Phillips theorem.

19.6. Prove that the function $u : [0, \infty) \to X$ defined by

$$u(t) = T(t)f, \quad t \geq 0,$$

is the unique solution of the initial value problem (19.1).

19.7. Let X be a complex Banach space. Let A be a closed linear operator from X into X with dense domain $\mathcal{D}(A)$. Prove that A is the infinitesimal generator of a one-parameter semigroup $\{T(t) : t \geq 0\}$ on X with

$$\|T(t)\| \leq 1, \quad t \geq 0,$$

if and only if

$$(0, \infty) \subseteq \rho(A)$$

and

$$\|R(\lambda, A)\| \leq \frac{1}{\lambda}.$$

for all $\lambda \in (0, \infty)$. (A one-parameter semigroup $\{T(t) : t \geq 0\}$ on a complex Banach space X with the property that

$$\|T(t)\| \leq 1, \quad t \geq 0,$$

is called a *one-parameter contraction semigroup*.)

19.8. Let X be a complex and separable Hilbert space in which the inner product is denoted by $(,)_X$. A closed linear operator from X into X with dense domain $\mathcal{D}(A)$ is said to be *dissipative* if

$$\mathrm{Re}\,(Ax, x)_X \leq 0, \quad x \in \mathcal{D}(A).$$

Prove that A is the infinitesimal generator of a one-parameter contraction semigroup if and only if A is dissipative and the operator $\lambda I - A$ is surjective for all $\lambda \in (0, \infty)$.

Chapter 20

Fredholm Operators

Among the basic questions in the study of equations are the existence and uniqueness of solutions. However, it is in the nature of the subject that we cannot always expect to have both for a given equation. So, we have to seek the second best. To see what is sensible to ask for, let us note that existence and uniqueness are intimately related to, respectively, the range and the null space of the operator associated with the equation under investigation. Intuitively, we naturally want the range to be *big* and the null space to be *small*. To make these ideas precise, we introduce Fredholm operators in this chapter. Since there are very lucid accounts of Fredholm operators in the literature, we give only the results that we need about these operators. Details and proofs can be found in [Schechter (2002)].

Definition 20.1. Let A be a bounded linear operator from a complex Banach space X into a complex Banach space Y. Suppose that
(i) the range $R(A)$ of A is a closed subspace of Y,
(ii) the null space $N(A)$ of A is finite-dimensional,
(iii) the null space $N(A^t)$ of the adjoint A^t of A is finite-dimensional.
Then we call A a *Fredholm operator*.

Definition 20.2. Let A be a Fredholm operator from X into Y as in Definition 20.1. Then the *index* $i(A)$ of A is defined by

$$i(A) = \dim N(A) - \dim N(A^t),$$

where $\dim N(A)$ and $\dim N(A^t)$ are the dimension of $N(A)$ and the dimension of $N(A^t)$ respectively.

Remark 20.3. The requirement that $N(A)$ be finite-dimensional is what we mean by saying that the null space of A is small. It is well known that the quotient space $Y/R(A)$ can be identified with $N(A^t)$. Thus, the condition that $N(A^t)$ is finite-dimensional can be translated into the requirement that the range $R(A)$ of A is big in Y.

Remark 20.4. Let K be a compact operator on X. Then it is well known that $I - K$ is a Fredholm operator on X and

$$i(I - K) = 0.$$

This fact is the Riesz theory of compact operators on Banach spaces.

The following theorem is the main tool that we use to study Fredholm pseudo-differential operators. It is known as Atkinson's theorem.

Theorem 20.5. (Atkinson's Theorem) *Let A be a bounded linear operator from X into Y. Then A is Fredholm if and only if we can find a bounded linear operator from Y into X, a compact operator K_1 on X and a compact operator K_2 on Y such that*

$$BA = I - K_1$$

on X, where I is the identity operator on X, and

$$AB = I - K_2$$

on Y, where I is the identity operator on Y.

We give a necessary and sufficient condition for a bounded linear operator to have a closed range.

Theorem 20.6. *Let A be a bounded linear operator from a complex Banach space X into a complex Banach space Y. Suppose that A is injective. Then the range $R(A)$ of A is closed in Y if and only if there exists a positive constant C such that*

$$\|x\|_X \leq C\|Ax\|_Y, \quad x \in X.$$

Proof Suppose that $R(A)$ is closed in Y. Then $R(A)$ is a Banach space with norm inherited from Y. So, $A : X \to R(A)$ is a bijective and bounded linear operator. Thus, by the bounded inverse theorem, $A^{-1} : R(A) \to X$ is a bounded linear operator. So, there exists a positive constant C such that

$$\|A^{-1}y\|_X \leq C\|y\|_Y, \quad y \in R(A).$$

This means that

$$\|A^{-1}Ax\|_X \leq C\|Ax\|_Y, \quad x \in X.$$

Thus,

$$\|x\|_X \leq C\|Ax\|_Y, \quad x \in X.$$

Conversely, suppose that there exists a positive constant C such that

$$\|x\|_X \leq C\|Ax\|_Y, \quad x \in X.$$

Let $\{y_k\}$ be a sequence in $R(A)$ such that

$$y_k \to y$$

in Y as $k \to \infty$. For $k = 1, 2, \ldots,$ let $x_k \in X$ be such that

$$Ax_k = y_k.$$

Then

$$\|x_j - x_k\|_X \leq C\|Ax_j - Ax_k\|_Y = \|y_j - y_k\|_Y \to 0$$

as $j, k \to \infty$. So, $\{x_k\}$ is a Cauchy sequence in X. Since X is complete, it follows that

$$x_k \to x$$

for some x in X as $k \to \infty$. But

$$Ax_k \to y$$

as $k \to \infty$. Therefore

$$Ax = y.$$

Hence $y \in R(A)$ and this proves that $R(A)$ is a closed subspace of Y. □

That the index behaves like the logarithm converting multipication into addition is made precise in the following theorem. Since the index is computed only for Fredholm operators on Hilbert spaces in this book, the theorem is stated and proved in a Hilbert space setting.

Theorem 20.7. *Let X, Y and Z be complex, separable and infinite-dimensional Hilbert spaces. Let $A_1 : X \to Y$ and $A_2 : Y \to Z$ be Fredholm operators. Then $A_2 A_1 : X \to Z$ is a Fredholm operator and*

$$i(A_2 A_1) = i(A_2) + i(A_1).$$

Before giving a proof of Theorem 20.7, it is helpful to give a formula for the null space $N(A^t)$ of the adjoint A^t of a Fredholm operator A from X into Y, where X and Y are complex and separable Hilbert spaces.

Proposition 20.8. *Let X and Y be complex and separable Hilbert spaces. Let A be a Fredholm operator from X into Y. Then*

$$Y = N(A^t) \oplus R(A).$$

The proof of Proposition 20.8 is left as an exercise. We note here that each of $N(A^t)$ and $R(A)$ is an orthogonal complement to the other. Of particular importance here is the fact that

$$N(A^t) = R(A)^{\perp},$$

which we also write as

$$N(A^t) = Y \ominus R(A).$$

Proof of Theorem 20.7 That $A_2 A_1 : X \to Z$ is a bounded linear operator is trivial. By Atkinson's Theorem 20.5, there exist bounded linear operators $B_1 : Y \to X$ and $B_2 : Z \to Y$, and compact operators K_1 on X, K_2 and K_3 on Y and K_4 on Z such that

$$B_1 A_1 = I - K_1$$

on X, where I is the identity operator on X,

$$A_1 B_1 = I - K_2$$

on Y, where I is the identity on Y,

$$B_2 A_2 = I - K_3$$

on Y, where I is the identity operator on Y and

$$A_2 B_2 = I - K_4$$

on Z, where I is the identity operator on Z. Thus,

$$B_1 B_2 A_2 A_1 = B_1 (I - K_3) A_1 = B_1 A_1 - B_1 K_3 A_1 = I - K_1 - B_1 K_3 A_1$$

on X and

$$A_2 A_1 B_1 B_2 = A_2 (I - K_2) B_2 = A_2 B_2 - A_2 K_2 B_2 = I - K_4 - A_2 K_2 B_2$$

on Z. Thus, by Atkinson's Theorem 20.5 again, $A_2 A_1 : X \to Z$ is Fredholm. Now, let M_1 be the finite-dimensional subspace of Y given by

$$M_1 = R(A_1) \cap N(A_2), \tag{20.1}$$

and we write

$$R(A_1) = M_1 \oplus M_2,$$

$$N(A_2) = M_1 \oplus M_3 \tag{20.2}$$

and

$$Y = R(A_1) \oplus M_3 \oplus M_4, \tag{20.3}$$

where M_3 and M_4 are finite-dimensional subspaces of Y and M_2 is a closed subspace of Y. Now, let X_1 be the subspace of X defined by

$$X_1 = N(A_2 A_1) \ominus N(A_1). \tag{20.4}$$

Then

$$A_1(X_1) = M_1. \tag{20.5}$$

Indeed, let $x_1 \in X_1$. Then $A_1 x_1 \in R(A_1)$. Also, by (20.4), $A_2 A_1 x_1 = 0$. Therefore, by (20.1), $A_1(X_1) \subseteq M_1$. Conversely, let $m_1 \in M_1$. Then, by (20.1), $m_1 = A_1 x$ for some $x \in X$ and, by (20.1) again,

$$A_2 m_1 = A_2 A_1 x = 0.$$

Therefore $x \in N(A_2 A_1)$. If we write $x = x_0 - w$, where $x_0 \in N(A_2 A_1)$ and $w \in N(A_1)$, then

$$A_1 x = A_1 x_0 - A_1 w = A_1 x = m_1.$$

Thus, $M_1 \subseteq A_1(X_1)$. Let Z_4 be the subspace of Z given by

$$Z_4 = R(A_2) \ominus R(A_2 A_1). \tag{20.6}$$

Then

$$Z_4 = A_2(M_4). \tag{20.7}$$

To prove (20.7), we just note that for all $y \in Y$, by (20.3),

$$A_2 y = A_2 A_1 x_0 + A_2 m_3 + A_2 m_4,$$

where $x_0 \in X$, $m_3 \in M_3$ and $m_4 \in M_4$. Thus,

$$A_2 y = A_2 A_1 x_0 + A_2 m_4,$$

and the proof of (20.7) is complete. Let $d_j = \dim M_j$, $j = 1, 3, 4$. Since $A_1 : X_1 \to M_1$ and $A_2 : M_4 \to Z_4$ are bijective, it follows from (20.5) and (20.7) that

$$\dim X_1 = d_1 \tag{20.8}$$

and

$$\dim Z_4 = d_4 \tag{20.9}$$

respectively. Thus, by (20.4)–(20.6),

$$\begin{aligned}
i(A_2 A_1) &= \dim N(A_2 A_1) - \dim \left(Z \ominus R(A_2 A_1) \right) \\
&= \dim N(A_1) + d_1 - \dim \left(Z \ominus R(A_2) \oplus Z_4 \right) \\
&= \dim N(A_1) + d_1 - \dim \left(Z \ominus R(A_2) \right) - d_4.
\end{aligned}$$

On the other hand, by (20.2) and (20.3),

$$i(A_2) + i(A_1)$$
$$= \dim N(A_2) - \dim (Z \ominus R(A_2)) + \dim N(A_1) - \dim (Y \ominus R(A_1))$$
$$= d_1 + d_3 - \dim (Z \ominus R(A_2)) + \dim N(A_1) - d_3 - d_4$$
$$= d_1 - \dim (Z \ominus R(A_2)) + \dim N(A_1) - d_4.$$

Thus,

$$i(A_2 A_1) = i(A_2) + i(A_1),$$

as claimed. □

Given a Fredholm operator, the problem of computing its index is usually an important but difficult problem. We give in this chapter a formula for computing the index of a Fredholm operator from a Hilbert space X into a Hilbert space Y. As the formula is expressed in terms of traces of trace class operators on Hilbert spaces, we first recall without proofs the basic facts on trace class operators and traces, which can be found in, for instance, [Lax (2002)] and [Reed and Simon (1980)].

Let A be a compact operator on a complex and separable Hilbert space X. Then $A^t A$ is a compact operator on X. Moreover, $A^t A$ is a *positive operator* on X in the sense that

$$(A^t A x, x) \geq 0, \quad x \in X.$$

Then it is well known that the operator $A^t A$ has a positive and compact square root, which we denote by $\sqrt{A^t A}$. This simply means that $\sqrt{A^t A}$ is a positive and compact operator on X such that its square is the same as $A^t A$. By the spectral theorem, there exists an orthonormal basis $\{\varphi_k\}$ for X consisting of eigenvectors of $\sqrt{A^t A}$. For $k = 1, 2, \ldots$, let s_k be the eigenvalue of $\sqrt{A^t A}$ corresponding to the eigenvector φ_k. Then we say that A is a *trace class operator* on X if

$$\sum_{k=1}^{\infty} s_k < \infty$$

and we say that A is a *Hilbert–Schmidt operator* on X if

$$\sum_{k=1}^{\infty} s_k^2 < \infty.$$

Remark 20.9. Let S_1 be the set of all trace class operators on X and let S_2 be the set of all Hilbert–Schmidt operators on X. Then it is well known that S_1 and S_2 are two-sided ideals in the set $B(X)$ of all bounded linear operators on X.

Let $A : X \to X$ be a bounded linear operator. Suppose that

$$\sum_{k=1}^{\infty} |(A\varphi_k, \varphi_k)_X| < \infty$$

for all orthonormal bases $\{\varphi_k\}$ for X. Then it can be proved that A is a trace class operator. Let A be a trace class operator on X. Then it can also be proved that $\sum_{k=1}^{\infty} (A\varphi_k, \varphi_k)_X$ is absolutely convergent for all orthonormal bases $\{\varphi_k\}$ for X and the sum is independent of the choice of the orthonormal basis. This sum is termed the *trace* of the trace class operator A on X and is denoted by $\mathrm{tr}(A)$. It is easy to prove that if A and B are trace class operators on X and c is any complex number, then $A + B$ and cA are also trace class operators on X. It is also easy to prove that

$$\mathrm{tr}(A + B) = \mathrm{tr}(A) + \mathrm{tr}(B)$$

and

$$\mathrm{tr}(cA) = c\,\mathrm{tr}(A).$$

We need the following property of the trace.

Theorem 20.10. *Let A be a trace class operator on a complex and separable Hilbert space X and let B be a bounded linear operator on X. Then AB and BA are also trace class operators on X. Moreover,*

$$\mathrm{tr}(AB) = \mathrm{tr}(BA).$$

A proof of Theorem 20.10 can be found on page 334 of [Lax (2002)]. As examples of Hilbert–Schmidt and trace class operators on $L^2(\mathbb{R}^n)$, we give one for Hilbert–Schmidt operators and one for trace class operators. They are useful to us for studying Fredholm pseudo-differential operators on $L^2(\mathbb{R}^n)$.

Theorem 20.11. *Let $A : L^2(\mathbb{R}^n) \to L^2(\mathbb{R}^n)$ be a bounded linear operator. Then A is a Hilbert–Schmidt operator if and only if there exists a function $K \in L^2(\mathbb{R}^n \times \mathbb{R}^n)$ such that*

$$(Af)(x) = \int_{\mathbb{R}^n} K(x, y)\, f(y)\, dy, \quad x \in \mathbb{R}^n,$$

for all $f \in L^2(\mathbb{R}^n)$.

The function K is usually called the *kernel* of the Hilbert–Schmidt operator A.

Theorem 20.12. *Let A be a bounded linear operator on $L^2(\mathbb{R}^n)$ given by*

$$(Af)(x) = \int_{\mathbb{R}^n} K(x,y)\, f(y)\, dy, \quad x \in \mathbb{R}^n,$$

for all $f \in L^2(\mathbb{R}^n)$, where K is a Schwartz function on $\mathbb{R}^n \times \mathbb{R}^n$. Then A is a trace class operator on $L^2(\mathbb{R}^n)$ and

$$\text{tr}(A) = \int_{\mathbb{R}^n} K(x,x)\, dx.$$

Theorem 20.12 is a special case of the more general results in [Brislawn (1988)].

We can now give the formula in Section 30.7 of [Lax (2002)] for the index of a Fredholm operator from one Hilbert space into another Hilbert space.

Theorem 20.13. *Let X and Y be complex, separable and infinite-dimensional Hilbert spaces. Let $A : X \to Y$ be a Fredholm operator. Suppose that we can find a bounded linear operator $B : Y \to X$, a trace class operator T_1 on X and a trace class operator T_2 on Y such that*

$$BA = I - T_1, \tag{20.10}$$

where I is the identity operator on X, and

$$AB = I - T_2, \tag{20.11}$$

where I is the identity operator on Y. Then the index $i(A)$ of A is given by

$$i(A) = \text{tr}(T_1) - \text{tr}(T_2).$$

Proof By (20.10) and (20.11), we get

$$ABA = A - AT_1$$

and

$$ABA = A - T_2A.$$

Therefore, upon subtraction, we get

$$AT_1 = T_2A. \tag{20.12}$$

Since A is Fredholm, we can write

$$X = N(A) \oplus X_0$$

and

$$Y = R(A) \oplus Y_0,$$

where X_0 and Y_0 are the orthogonal complements of $N(A)$ in X and $R(A)$ in Y respectively. Let P be the projection of X onto X_0 along $N(A)$. Then

$$AP = A. \qquad (20.13)$$

Indeed, let $x \in X$. Then we can write $x = z + x_0$, where $z \in N(A)$ and $x_0 \in X_0$. So,

$$APx = Ax_0$$

and

$$Ax = Ax_0.$$

Thus, $APx = Ax$, which gives (20.13). By (20.12) and (20.13), we get

$$APT_1 = T_2 A. \qquad (20.14)$$

It is obvious that $A : X_0 \to R(A)$ is a bijection. Since $P : X \to X_0$ is a projection, it follows that PT_1 maps X_0 into X_0. By (20.12), T_2 maps $R(A)$ into $R(A)$. Now,

$$\operatorname{tr}\left(PT_1|_{X_0}\right) = \operatorname{tr}\left(T_2|_{R(A)}\right). \qquad (20.15)$$

Indeed, by Exercise 20.11, there exists a surjective isometry U of $R(A)$ onto X_0. Then, by (20.14),

$$UAPT_1 = UT_2 A = UT_2 U^{-1} UA$$

and hence

$$(UA)(PT_1)(UA)^{-1} = UT_2 U^{-1}.$$

Since each factor on the left hand side and the operator on the right hand side of the preceding equation maps X_0 into X_0, we can invoke Theorem 20.10 to conclude that

$$\operatorname{tr}\left(PT_1|_{X_0}\right) = \operatorname{tr}\left(UT_2 U^{-1}|_{X_0}\right). \qquad (20.16)$$

Now, we note that

$$\operatorname{tr}\left(UT_2 U^{-1}|_{X_0}\right) = \operatorname{tr}\left(T_2|_{R(A)}\right). \qquad (20.17)$$

Indeed, let $\{x_j^0\}$ be an orthonormal basis for X_0. Then $\{U^{-1}x_j^0\}$ is an orthonormal basis for $R(A)$. Thus,

$$\operatorname{tr}\left(UT_2 U^{-1}|_{X_0}\right) = \sum_{j=1}^{\infty}(UT_2 U^{-1}x_j^0, x_j^0)_X$$

$$= \sum_{j=1}^{\infty}(T_2 U^{-1}x_j^0, U^{-1}x_j^0)_Y = \operatorname{tr}\left(T_2|_{R(A)}\right).$$

So, (20.17) is proved and (20.15) follows from (20.16) and (20.17). Let $\{z_j\}_{j=1}^{\dim N(A)}$ be an orthonormal basis for $N(A)$ and let $\{x_j^0\}$ be an orthonormal basis for X_0. By (20.10), $T_1 = I$ on $N(A)$. It follows that

$$
\begin{aligned}
\mathrm{tr}(T_1) &= \sum_{j=1}^{\dim N(A)} (T_1 z_j, z_j)_X + \sum_{j=1}^{\infty} (T_1 x_j^0, x_j^0)_X \\
&= \sum_{j=1}^{\dim N(A)} (z_j, z_j)_X + \sum_{j=1}^{\infty} (T_1 x_j^0, P x_j^0)_X \\
&= \sum_{j=1}^{\dim N(A)} 1 + \sum_{j=1}^{\infty} (P T_1 x_j^0, x_j^0)_X \\
&= \dim N(A) + \mathrm{tr}\left(P T_1 |_{X_0} \right).
\end{aligned}
\tag{20.18}
$$

Let $\{y_j^0\}_{j=1}^{\dim N(A^t)}$ be an orthonormal basis for Y_0 and let $\{w_j\}$ be an orthonormal basis for $R(A)$. Then

$$
\begin{aligned}
\mathrm{tr}(T_2) &= \sum_{j=1}^{\dim N(A^t)} (T_2 y_j^0, y_j^0)_Y + \sum_{j=1}^{\infty} (T_2 w_j, w_j)_Y \\
&= \sum_{j=1}^{\dim N(A^t)} (T_2 y_j^0, y_j^0)_Y + \mathrm{tr}\left(T_2 |_{R(A)} \right).
\end{aligned}
\tag{20.19}
$$

Since $R(I - T_2) \subseteq R(A)$, we see that

$$
R(I - T_2) \perp Y_0.
$$

Therefore

$$
((I - T_2) y_j^0, y_j^0)_Y = 0, \quad j = 1, 2, \ldots.
$$

So, by (20.19),

$$
\begin{aligned}
\mathrm{tr}(T_2) &= \sum_{j=1}^{\dim N(A^t)} (y_j^0, y_j^0)_Y + \mathrm{tr}\left(T_2 |_{R(A)} \right) \\
&= \sum_{j=1}^{\dim N(A^t)} 1 + \mathrm{tr}\left(T_2 |_{R(A)} \right) \\
&= \dim N(A^t) + \mathrm{tr}\left(T_2 |_{R(A)} \right).
\end{aligned}
\tag{20.20}
$$

Thus, by (20.15), (20.18) and (20.20),

$$
i(A) = \dim N(A) - \dim N(A^t) = \mathrm{tr}(T_1) - \mathrm{tr}(T_2).
$$

$\square$

Exercises

20.1. Can a compact operator from a complex Banach space X into a complex Banach space Y be Fredholm? Explain your answer.

20.2. Can a Fredholm operator from a complex Banach space X into a complex Banach space Y be compact? Explain your answer.

20.3. Let X and Y be complex, separable and infinite-dimensional Hilbert spaces. Let $A : X \to Y$ be a Fredholm operator and let $K : X \to Y$ be a compact operator. Prove that $A + K : X \to Y$ is Fredholm and

$$i(A + K) = i(A).$$

20.4. Let A be a bounded linear operator on a complex and separable Hilbert space X. Prove that $A^t A$ is a positive operator on X.

20.5. Let A be a bounded and positive operator on a complex and separable Hilbert space X. Prove that A is self-adjoint, i.e., $A^t = A$.

20.6. Let $A : L^2(\mathbb{R}^n) \to L^2(\mathbb{R}^n)$ be a Hilbert–Schimdt operator given by the kernel K. Prove that the kernel K^t of $A^t : L^2(\mathbb{R}^n) \to L^2(\mathbb{R}^n)$ is given by

$$K^t(x,y) = \overline{K(y,x)}, \quad x,y \in \mathbb{R}^n.$$

20.7. Prove that S_1 is a subspace of $B(X)$.

20.8. Prove that $\mathrm{tr} : S_1 \to \mathbb{C}$ is a linear functional.

20.9. Prove that if A is a Fredholm operator on a complex ansd separable Hilbert space X, then

$$X = N(A^t) \oplus R(A).$$

(This exercise says that $N(A^t)$ and $R(A)$ are orthogonal to each other.)

20.10. Let M be a closed subspace of a complex and separable Hilbert space X. If we write $X = M \oplus M^\perp$ and let P be the projection of X onto M along $M^\perp$. Find P^t.

20.11. Find a surjective isometry $U : R(A_1) \to X_0$ used in the proof of Theorem 20.13

Chapter 21

Fredholm Pseudo-Differential Operators

We begin with proving that Fredholm pseudo-differential operators on $L^p(\mathbb{R}^n)$, $1 < p < \infty$, with symbols in S^m, $-\infty < m < \infty$, are elliptic. To this end, we need some technical preparations.

Definition 21.1. For $\lambda > 0$, $\tau \geq 0$ and $x_0, \xi_0 \in \mathbb{R}^n$, we define the operator $R_{\lambda,\tau}(x_0, \xi_0) : L^p(\mathbb{R}^n) \to L^p(\mathbb{R}^n)$, $1 < p < \infty$, by

$$(R_{\lambda,\tau}(x_0, \xi_0)u)(x) = \lambda^{\tau n/p} e^{i\lambda x \cdot \xi_0} u(\lambda^\tau(x - x_0)), \quad x \in \mathbb{R}^n,$$

for all $u \in L^p(\mathbb{R}^n)$.

Proposition 21.2. The operator $R_{\lambda,\tau}(x_0, \xi_0) : L^p(\mathbb{R}^n) \to L^p(\mathbb{R}^n)$ is a surjective isometry and the inverse is given by

$$(R_{\lambda,\tau}(x_0, \xi_0)^{-1}u)(x) = \lambda^{-\tau n/p} e^{-i\lambda(x_0 + \lambda^{-\tau}x) \cdot \xi_0} u(x_0 + \lambda^{-\tau}x), \quad x \in \mathbb{R}^n,$$

for all $u \in L^p(\mathbb{R}^n)$.

Proof We get

$$\|R_{\lambda,\tau}(x_0, \xi_0)u\|_p^p$$

$$= \int_{\mathbb{R}^n} |(R_{\lambda,\tau}(x_0, \xi_0)u)(x)|^p \, dx$$

$$= \int_{\mathbb{R}^n} \lambda^{\tau n} |u(\lambda^\tau(x - x_0))|^p dx, \quad u \in L^p(\mathbb{R}^n). \tag{21.1}$$

Let $y = \lambda^\tau(x - x_0)$ in (21.1). Then we have

$$\|R_{\lambda,\tau}(x_0, \xi_0)u\|_p = \|u\|_p, \quad u \in L^p(\mathbb{R}^n).$$

This proves that $R_{\lambda,\tau}(x_0, \xi_0)$ is an isometry. Now, let $v \in L^p(\mathbb{R}^n)$. We need to find a function u in $L^p(\mathbb{R}^n)$ such that

$$R_{\lambda,\tau}(x_0, \xi_0)u = v.$$

Let u be the function in $L^p(\mathbb{R}^n)$ given by

$$u(x) = \lambda^{-\tau n/p} e^{-i\lambda(x_0 + \lambda^{-\tau}x)\cdot\xi_0} v(x_0 + \lambda^{-\tau}x), \quad x \in \mathbb{R}^n.$$

Then

$$R_{\lambda,\tau}(x_0, \xi_0)u = v.$$

$\square$

Proposition 21.3. *For all $u \in L^p(\mathbb{R}^n)$ and $v \in L^{p'}(\mathbb{R}^n)$, where $\frac{1}{p} + \frac{1}{p'} = 1$, $1 < p < \infty$,*

$$(R_{\lambda,\tau}(x_0, \xi_0)u, v) \to 0$$

as $\lambda \to \infty$.

Proof Let $u, v \in C_0^\infty(\mathbb{R}^n)$. Then we have

$$|(R_{\lambda,\tau}(x_0, \xi_0)u, v)|$$

$$\leq \lambda^{\tau n/p} \int_{\mathbb{R}^n} |u(\lambda^\tau(x - x_0))||v(x)|\, dx$$

$$= \lambda^{\tau n/p} \int_{\mathbb{R}^n} \lambda^{-\tau n} |u(y)||v(x_0 + \lambda^{-\tau}y)|\, dy$$

$$= \lambda^{-\tau n/p'} \int_{\mathbb{R}^n} |v(x_0 + \lambda^{-\tau}y)||u(y)|\, dy$$

$$\leq \lambda^{-\tau n/p'} \sup_{y \in \mathbb{R}^n} |v(y)| \int_{\mathbb{R}^n} |u(y)|\, dy. \tag{21.2}$$

So, by (21.2), $(R_{\lambda,\tau(x_0,\xi_0)}u, v) \to 0$ as $\lambda \to \infty$ for all $u, v \in C_0^\infty(\mathbb{R}^n)$. Let $u \in L^p(\mathbb{R}^n)$ and $v \in L^{p'}(\mathbb{R}^n)$. By density, we can find sequences $\{\varphi_j\}$ and $\{\psi_j\}$ in $C_0^\infty(\mathbb{R}^n)$ such that

$$\varphi_j \to u$$

and

$$\psi_j \to v$$

in $L^p(\mathbb{R}^n)$ and $L^{p'}(\mathbb{R}^n)$ as $j \to \infty$ respectively. Therefore for every positive number ε, there exists a positive integer J such that

$$|(R_{\lambda,\tau}(x_0, \xi_0)u, v) - (R_{\lambda,\tau}(x_0, \xi_0)u_j, v_j)| < \varepsilon$$

for all $j \geq J$. Then, again using the fact that $(R_{\lambda,\tau}(x_0, \xi_0)u_j, v_j) \to 0$ as $\lambda \to \infty$ for $u_j, v_j \in C_0^\infty(\mathbb{R}^n)$, the result follows immediately. $\square$

Proposition 21.4. *Let $\sigma \in S^m$. Then*

$$R_{\lambda,\tau}(x_0, \xi_0)^{-1} T_\sigma R_{\lambda,\tau}(x_0, \xi_0) = T_{\sigma_{\lambda,\tau}}, \tag{21.3}$$

where

$$\sigma_{\lambda,\tau}(x,\eta) = \sigma(x_0 + \lambda^{-\tau}x, \lambda\xi_0 + \lambda^{\tau}\eta), \quad x,\eta \in \mathbb{R}^n. \tag{21.4}$$

Moreover, if $\sigma \in S^0$, $\lambda \geq 1$, $0 \leq \tau \leq 1/2$ *and* $\xi_0 \neq 0$, *then for all multi-indices* α *and* β, *there exists a positive constant* C_β *such that*

$$|(\partial_x^\alpha \partial_\eta^\beta \sigma_{\lambda,\tau})(x,\eta)| \leq C_\beta p_{\alpha,\beta}(\sigma) \frac{\langle\eta\rangle^{|\beta|}}{|\xi_0|^{|\beta|}} \lambda^{-\tau|\alpha|}\lambda^{-(1-2\tau)|\beta|}, \quad x,\eta \in \mathbb{R}^n, \tag{21.5}$$

where $p_{\alpha,\beta}$ *denotes the corresponding norm in* S^0.

An ingredient in the proof of Proposition 21.4 is the following *Peetre's inequality*.

Lemma 21.5. (Peetre's Inequality) *For all* $t \in (-\infty, \infty)$ *and for all* $x, y \in \mathbb{R}^n$,

$$\left(\frac{1 + |x|^2}{1 + |y|^2}\right)^t \leq 2^{|t|}(1 + |x - y|^2)^{|t|}.$$

Proof The inequality is obviously true if $t = 0$. For all y and z in $\mathbb{R}^n$,

$$\begin{aligned}
1 + |y - z|^2 &= 1 + (y - z) \cdot (y - z) \\
&= 1 + |y|^2 - 2y \cdot z + |z|^2 \\
&\leq 1 + |y|^2 + 2|y|\,|z| + |z|^2.
\end{aligned}$$

Since

$$2|y|\,|z| \leq |y|^2 + |z|^2,$$

it follows that

$$1 + |y - z|^2 \leq 1 + 2|y|^2 + 2|z|^2 \leq 2(1 + |y|^2)(1 + |z|^2)$$

for all y and z in $\mathbb{R}^n$. Replacing z by $y - x$, we get

$$1 + |x|^2 \leq 2(1 + |y|^2)(1 + |x - y|^2).$$

If $t > 0$, then

$$(1 + |x|^2)^t \leq 2^t(1 + |y|^2)^t(1 + |x - y|^2)^t,$$

as asserted. If $t < 0$, then $-t > 0$ and by what we have proved,

$$(1 + |y|^2)^{-t} \leq 2^{-t}(1 + |x|^2)^{-t}(1 + |x - y|^2)^{-t},$$

which is the same as

$$(1 + |x|^2)^t \leq 2^{|t|}(1 + |y|^2)^t(1 + |x - y|^2)^{|t|},$$

as required.   □

Proof of Proposition 21.4 We first note that for all $\varphi \in \mathcal{S}$,

$$
\begin{aligned}
(T_\sigma \varphi)(x) &= (2\pi)^{-n/2} \int_{\mathbb{R}^n} e^{ix\cdot\xi} \sigma(x,\xi) \hat\varphi(\xi)\, d\xi \\
&= (2\pi)^{-n} \int_{\mathbb{R}^n} \int_{\mathbb{R}^n} e^{i(x-y)\cdot\xi} \sigma(x,\xi) u(y)\, dy\, d\xi,
\end{aligned}
$$

where the integral $\int_{\mathbb{R}^n} \int_{\mathbb{R}^n}$ is understood to be an iterated integral in which the integration is first performed with respect to y and then ξ. Let $u \in \mathcal{S}$. Then for all $x \in \mathbb{R}^n$,

$$
\begin{aligned}
(R_{\lambda,\tau}&(x_0,\xi_0)^{-1} T_\sigma R_{\lambda,\tau}(x_0,\xi_0) u)(x) \\
&= e^{-i\lambda(x_0+\lambda^{-\tau}x)\cdot\xi_0}(2\pi)^{-n} \int_{\mathbb{R}^n} \int_{\mathbb{R}^n} e^{i\lambda(x_0+\lambda^{-\tau}x-y)\cdot\xi}\sigma(x_0+\lambda^{-\tau}x,\xi) \\
&\qquad\qquad\qquad\qquad\qquad\qquad e^{i\lambda y\cdot\xi_0} u(\lambda^\tau(y-x_0))\, dy\, d\xi \\
&= \lambda^{-\tau n}(2\pi)^{-n} \int_{\mathbb{R}^n} \int_{\mathbb{R}^n} e^{i\lambda^{-\tau}(\xi-\lambda\xi_0)\cdot(x-z)}\sigma(x_0+\lambda^{-\tau}x,\xi) u(z)\, dz\, d\xi \\
&= (2\pi)^{-n} \int_{\mathbb{R}^n} \int_{\mathbb{R}^n} e^{i(x-z)\cdot\eta}\sigma(x_0+\lambda^{-\tau}x,\lambda\xi_0+\lambda^\tau\eta) u(z)\, dz\, d\eta.
\end{aligned}
$$

Thus, we get (21.3) and (21.4), as asserted. Now, using (21.4), the chain rule and Peetre's inequality,

$$
\begin{aligned}
|(\partial_x^\alpha \partial_\eta^\beta \sigma_{\lambda,\tau})(x,\eta)| &= |(\partial_x^\alpha \partial_\eta^\beta \sigma)(x_0+\lambda^{-\tau}x,\lambda\xi_0+\lambda^\tau\eta)\lambda^{-\tau|\alpha|}\lambda^{\tau|\beta|}| \\
&\leq p_{\alpha,\beta}(\sigma)(1+|\lambda\xi_0+\lambda^\tau\eta|)^{-|\beta|}\lambda^{-\tau|\alpha|}\lambda^{\tau|\beta|} \\
&\leq C_\beta p_{\alpha,\beta}(\sigma)\langle\lambda\xi_0+\lambda^\tau\eta\rangle^{-|\beta|}\lambda^{-\tau|\alpha|}\lambda^{\tau|\beta|} \\
&\leq C_\beta p_{\alpha,\beta}(\sigma)\langle\lambda\xi_0\rangle^{-|\beta|}\langle\lambda^\tau\eta\rangle^{|\beta|}\lambda^{-\tau|\alpha|} \\
&\leq C_\beta p_{\alpha,\beta}(\sigma)|\xi_0|^{-|\beta|}\langle\eta\rangle^{|\beta|}\lambda^{(2\tau-1)|\beta|}\lambda^{-\tau|\alpha|}. \qquad (21.6)
\end{aligned}
$$

Hence (21.5) is proved.   □

The following theorem is one of the main theorems of this chapter.

Theorem 21.6. *Let $\sigma \in S^0$ be such that $T_\sigma : L^p(\mathbb{R}^n) \to L^p(\mathbb{R}^n)$ is a Fredholm operator for $1 < p < \infty$. Then there are positive constants C and R such that*

$$
|\sigma(x,\xi)| \geq C
$$

for all $x \in \mathbb{R}^n$ and $|\xi| \geq R$, i.e., σ is elliptic.

Proof Since T_σ is a Fredholm operator, it follows that we can find a bounded linear operator S on $L^p(\mathbb{R}^n)$ and a compact operator K on $L^p(\mathbb{R}^n)$ such that

$$ST_\sigma = I + K.$$

Let M be the set of all points ξ in $\mathbb{R}^n$ such that there exists a point x in $\mathbb{R}^n$ for which

$$|\sigma(x,\xi)| \leq \frac{1}{2\|S\|}.$$

Now, if M is bounded, then there exists a positive number R such that

$$|\xi| \leq R, \quad \xi \in M.$$

Thus, for each point $\xi \in \mathbb{R}^n$ with $|\xi| \geq R$, we get, for all $x \in \mathbb{R}^n$,

$$|\sigma(x,\xi)| \geq \frac{1}{2\|S\|},$$

which is the same as saying that σ is elliptic. So, suppose that M is not bounded. Then there exists a sequence $\{(x_k,\xi_k)\}$ in $\mathbb{R}^n \times \mathbb{R}^n$ such that

$$|\xi_k| \to \infty$$

as $k \to \infty$ and

$$|\sigma(x_k,\xi_k)| \leq \frac{1}{2\|S\|}, \quad k = 1, 2, \ldots.$$

Thus, there exists a subsequence of $\{(x_k,\xi_k)\}$, again denoted by $\{(x_k,\xi_k)\}$, such that

$$\sigma(x_k,\xi_k) \to \sigma_\infty$$

for some complex number σ_∞ as $k \to \infty$. Therefore

$$|\sigma_\infty| \leq \frac{1}{2\|S\|}. \tag{21.7}$$

For $k = 1, 2, \ldots$, let $\lambda_k = |\xi_k|$. Then, by Proposition 21.4, we have

$$R_{\lambda_k,\tau}\left(x_k, \frac{\xi_k}{|\xi_k|}\right)^{-1} T_\sigma R_{\lambda_k,\tau}\left(x_k, \frac{\xi_k}{|\xi_k|}\right) = T_{\sigma_{\lambda_k,\tau}},$$

where

$$\sigma_{\lambda_k,\tau}(x,\eta) = \sigma(x_k + \lambda_k^{-\tau}x, \xi_k + \lambda_k^\tau \eta), \quad x, \eta \in \mathbb{R}^n.$$

Let α and β be multi-indices. Then, in view of (21.5), there exists a positive constant C_β such that

$$|(\partial_x^\alpha \partial_\eta^\beta \sigma_{\lambda_k,\tau})(x,\eta)| \leq C_\beta p_{\alpha,\beta}(\sigma)\langle\eta\rangle^{|\beta|}\lambda_k^{-\tau|\alpha|}\lambda_k^{-(1-2\tau)|\beta|}. \tag{21.8}$$

For $k = 1, 2, \ldots$, let σ_k^∞ be given by

$$\sigma_k^\infty = \sigma_{\lambda_k,\tau}(0,0) = \sigma(x_k, \xi_k).$$

Using Taylor's formula given in Theorem 7.3 and the estimate (21.8), we get

$$
\begin{aligned}
&|\sigma_{\lambda_k,\tau}(x,\eta) - \sigma_k^\infty| \\
&= |\sigma_{\lambda_k,\tau}(x,\eta) - \sigma_{\lambda_k,\tau}(0,0)| \\
&= \left| \sum_{|\gamma+\mu|=1} x^\gamma \eta^\mu \int_0^1 (\partial_x^\gamma \partial_\eta^\mu \sigma_{\lambda_k,\tau})(\theta x, \theta \eta)\, d\theta \right| \\
&= \sum_{|\gamma+\mu|=1} |x|^{|\gamma|} |\eta|^{|\mu|} \int_0^1 C_\mu p_{\gamma,\mu}(\sigma) \langle \eta \rangle^{|\mu|} \lambda_k^{-\tau|\gamma|} \lambda_k^{-(1-2\tau)|\mu|}\, d\theta \to 0
\end{aligned}
$$

$$(21.9)$$

uniformly for (x,η) on every compact subset K of $\mathbb{R}^n \times \mathbb{R}^n$ as $k \to \infty$. Let $u \in \mathcal{S}$. Then

$$(T_{\sigma_{\lambda_k,\tau}} u)(x) - \sigma_k^\infty u(x) = (2\pi)^{-n/2} \int_{\mathbb{R}^n} e^{ix\cdot\eta} (\sigma_{\lambda_k,\tau}(x,\eta) - \sigma_k^\infty) \hat{u}(\eta)\, d\eta$$

for all $x \in \mathbb{R}^n$. By (21.9), the assumption that $\sigma \in S^0$ and Lebesgue's dominated convergence theorem,

$$(T_{\sigma_{\lambda_k,\tau}} u)(x) \to \sigma_k^\infty u(x) \tag{21.10}$$

for all $x \in \mathbb{R}^n$ as $k \to \infty$. Moreover, for all $l \in \mathbb{N}$, using (21.8) and an integration by parts, we can find a positive constant C_μ for each μ with $|\mu| \le 2l$ such that

$$
\begin{aligned}
&|\langle x \rangle^{2l} (T_{\sigma_{\lambda_k,\tau}} u)(x)| \\
&= \left| \langle x \rangle^{2l} (2\pi)^{-n/2} \int_{\mathbb{R}^n} e^{ix\cdot\eta} \sigma_{\lambda_k,\tau}(x,\eta) \hat{u}(\eta)\, d\eta \right| \\
&= \left| (2\pi)^{-n/2} \int_{\mathbb{R}^n} ((I - \Delta_\xi)^l e^{ix\cdot\xi}) \sigma_{\lambda_k,\tau}(x,\xi) \hat{u}(\xi)\, d\xi \right| \\
&= (2\pi)^{-n/2} \int_{\mathbb{R}^n} \left| \sum_{|\mu| \le 2l} \frac{1}{\mu!} (P^\mu(D)\hat{u})(\xi) \right| C_\mu \langle \xi \rangle^{|\mu|}\, d\xi
\end{aligned}
$$

for all $x \in \mathbb{R}^n$, where $P(D) = (I - \Delta)^l$. So, there exists a positive constant C such that

$$|(T_{\sigma_{\lambda_k,\tau}} u)(x)|^p \le C \langle x \rangle^{-2lp}, \quad x \in \mathbb{R}^n.$$

Now, if $2lp > n$, then $\langle x \rangle^{-2lp} \in L^1(\mathbb{R}^n)$. So, there exists a positive constant C_1 such that

$$|(T_{\sigma_{\lambda_k},\tau} - \sigma_k^\infty)u)(x)| \leq C_1 \langle x \rangle^{-2lp}, \quad x \in \mathbb{R}^n. \tag{21.11}$$

Thus,

$$T_{\sigma_{\lambda_k},\tau} u \to \sigma_\infty u$$

in $L^p(\mathbb{R}^n)$ as $k \to \infty$. Let u be a nonzero function $L^p(\mathbb{R}^n)$. Since $R_{\lambda_k,\tau}\left(x_k, \frac{\xi_k}{|\xi_k|}\right)$ is an isometry, it follows that

$$
\begin{aligned}
0 < \|u\|_p &= \left\| R_{\lambda_k,\tau}\left(x_k, \frac{\xi_k}{|\xi_k|}\right)u \right\|_p \\
&= \left\| (ST_\sigma - K)R_{\lambda_k,\tau}\left(x_k, \frac{\xi_k}{|\xi_k|}\right)u \right\|_p \\
&\leq \left\| ST_\sigma R_{\lambda_k,\tau}\left(x_k, \frac{\xi_k}{|\xi_k|}\right)u \right\|_p + \left\| KR_{\lambda_k,\tau}\left(x_k, \frac{\xi_k}{|\xi_k|}\right)u \right\|_p \\
&\leq \left\| R_{\lambda_k,\tau}\left(x_k, \frac{\xi_k}{|\xi_k|}\right)^{-1} T_\sigma R_{\lambda_k,\tau}\left(x_k, \frac{\xi_k}{|\xi_k|}\right)u \right\|_p \|S\| \\
&\quad + \left\| KR_{\lambda_k,\tau}\left(x_k, \frac{\xi_k}{|\xi_k|}\right)u \right\|_p . \tag{21.12}
\end{aligned}
$$

Now, using the fact that K is a compact operator and Proposition 21.3, it follows that

$$\left\| KR_{\lambda_k,\tau}\left(x_k, \frac{\xi_k}{|\xi_k|}\right)u \right\|_p \to 0$$

as $k \to \infty$. Then by (21.12),

$$\|u\|_p \leq \|S\| |\sigma_\infty| \|u\|_p. \tag{21.13}$$

Then (21.7) and (21.13) give the contradiction that

$$\frac{1}{\|S\|} \leq |\sigma_\infty| \leq \frac{1}{2\|S\|},$$

which completes the proof. $\square$

The preceding theorem can be generalized to the following theorem.

Theorem 21.7. *Let $\sigma \in S^m$, $-\infty < m < \infty$, and let $T_\sigma : H^{s,p} \to H^{s-m,p}$ be a Fredholm operator for some $s \in (-\infty, \infty)$. Then T_σ is an elliptic operator.*

Proof The operators $T_\sigma : H^{s,p} \to H^{s-m,p}$, $J_{-s} : H^{s,p} \to L^p(\mathbb{R}^n)$ and $J_{m-s} : H^{s-m,p} \to L^p(\mathbb{R}^n)$ are bounded linear operators. Here J_s, the Bessel potential, is a pseudo-differential operator with symbol $\sigma_s \in S^{-s}$, where

$$\sigma_s(\xi) = \langle \xi \rangle^{-s}, \quad \xi \in \mathbb{R}^n.$$

Let

$$J_{m-s} T_\sigma J_s = T_\tau.$$

Then

$$T_\tau : L^p(\mathbb{R}^n) \to L^p(\mathbb{R}^n),$$

where $\tau \in S^0$. Since J_s is bijective, it follows that J_s is Fredholm and elliptic for all $s \in (-\infty, \infty)$. So, T_τ is elliptic. By the fact that J_s, $s \in \mathbb{R}^n$, is bijective, it follows immediately that T_σ is elliptic. $\qquad \square$

For pseudo-differential operators with symbols in S^m, $-\infty < m < \infty$, that ellipticity does not imply Fredholmness can be seen from the following example.

Example 21.8. Let $P(D) = \sum_{|\alpha| \leq m} a_\alpha D^\alpha$ be a linear partial differential operator with constant coefficients on $\mathbb{R}^n$ such that $P(D) : H^{m,p} \to L^p(\mathbb{R}^n)$ is injective and there exists a sequence $\{\xi_k\}$ in $\mathbb{R}^n$ such that

$$P(\xi_k) \to 0$$

as $k \to \infty$. Then $P(D) : H^{m,p} \to L^p(\mathbb{R}^n)$ is not Fredholm. More precisely, the range $R(P)$ of the operator $P(D) : H^{m,p} \to L^p(\mathbb{R}^n)$ is not a closed subspace of $L^p(\mathbb{R}^n)$. To this end, let φ be a function in $\mathcal{S}$ such that $\|\varphi\|_p = 1$. Let $\{\varepsilon_k\}$ be a sequence of positive numbers such that

$$\varepsilon_k^{|\mu|} P^{(\mu)}(\xi_k) \to 0$$

for $0 < |\mu| \leq m$ as $k \to \infty$. Let $\{\varphi_k\}$ be the sequence of functions in $\mathcal{S}$ defined by

$$\varphi_k(x) = \varepsilon_k^{-n/p} \varphi(\varepsilon_k x) e^{ix \cdot \xi_k}, \quad x \in \mathbb{R}^n,$$

for $k = 1, 2, \ldots$. Then

$$\|\varphi_k\|_p = 1, \quad k = 1, 2, \ldots.$$

By Leibniz's formula, we get for $k = 1, 2, \ldots,$

$$(P(D)\varphi_k)(x)$$

$$= \varepsilon_k^{-n/p} \sum_{|\mu| \leq m} \frac{1}{\mu!} (P^{(\mu)}(D)e^{ix \cdot \xi_k})(D^\mu \varphi)(\varepsilon_k x)\varepsilon_k^{|\mu|}$$

$$= \varepsilon_k^{-n/p} \sum_{|\mu| \leq m} \frac{1}{\mu!} P^{(\mu)}(\xi_k)e^{ix \cdot \xi_k}(D^\mu \varphi)(\varepsilon_k x)\varepsilon_k^{|\mu|}$$

$$= \varepsilon_k^{-n/p} P(\xi_k)e^{ix \cdot \xi_k}\varphi(\varepsilon_k x) + \varepsilon_k^{-n/p} \sum_{1 \leq |\mu| \leq m} \frac{1}{\mu!} P^{(\mu)}(\xi_k)e^{ix \cdot \xi_k}(D^\mu \varphi)(\varepsilon_k x)\varepsilon_k^{|\mu|}$$

for all $x \in \mathbb{R}^n$. So, for $k = 1, 2, \ldots,$

$$\|P(D)\varphi_k\|_p \leq |P(\xi_k)| + \sum_{1 \leq |\mu| \leq m} \frac{1}{\mu!} \varepsilon_k^{|\mu|} |P^\mu(\xi_k)| \, \|D^\mu \varphi\|_p \to 0$$

as $k \to \infty$. Thus, we cannot find a positive constant C such that

$$\|\varphi_k\|_p \leq C\|P(D)\varphi_k\|_p, \quad k = 1, 2, \ldots.$$

So, the range $R(P)$ is not closed in $L^p(\mathbb{R}^n)$ by Theorem 20.6.

For a positive result, we have the following proposition.

Proposition 21.9. *Let $P(D)$ be a linear partial differential operator with constant coefficients and of order m on $\mathbb{R}^n$. Then $P(D) : H^{m,2} \to L^2(\mathbb{R}^n)$ is Fredholm with zero index if and only if*

$$0 \notin \overline{\{P(\xi) : \xi \in \mathbb{R}^n\}},$$

where $\overline{\{\cdots\}}$ is the closure in $\mathbb{R}^n$ of the set $\{\cdots\}$.

Before giving a proof of Proposition 21.9, we need two lemmas.

Lemma 21.10. *Let P be a polynomial on $\mathbb{R}^n$ given by*

$$P(\xi) = \sum_{|\alpha| \leq m} a_\alpha \xi^\alpha, \quad \xi \in \mathbb{R}^n.$$

Let $Z(P)$ be the zero set of P, i.e.,

$$Z(P) = \{\xi \in \mathbb{R}^n : P(\xi) = 0\}.$$

Then $m(Z(P)) = 0$, where m is the Lebesgue measure on $\mathbb{R}^n$.

Proof The lemma is certainly true for $n = 1$. Suppose that the lemma is true for all polynomials on $\mathbb{R}^{n-1}$. Write

$$\mathbb{R}^n = \mathbb{R}^{n-1} \times \mathbb{R}$$

and any point in $\mathbb{R}^n$ is of the form (ξ', ξ_n), where $\xi' \in \mathbb{R}^{n-1}$. Any multi-index $\alpha = (\alpha_1, \alpha_2, \ldots, \alpha_n)$ can be written as (α', α_n), where

$$\alpha' = (\alpha_1, \alpha_2, \ldots, \alpha_{n-1}).$$

Now,

$$P(\xi) = P(\xi', \xi_n) = \sum_{|\alpha'| \leq m - \alpha_n} \left(\sum_{\alpha_n = 0}^{m} a_{\alpha', \alpha_n} \xi_n^{\alpha_n} \right) \xi'^{\alpha'}.$$

Thus,

$$Z(P) = \left\{ (\xi', \xi_n) \in \mathbb{R}^n : \sum_{|\alpha'| \leq m - \alpha_n} \left(\sum_{\alpha_n = 0}^{m} a_{\alpha', \alpha_n} \xi_n^{\alpha_n} \right) \xi'^{\alpha'} = 0 \right\}.$$

Let $\chi_{Z(P)}$ be the characteristic function of $Z(P)$. Then, by the induction hypothesis, we get

$$m(Z(P)) = \int_{-\infty}^{\infty} \int_{\mathbb{R}^{n-1}} \chi_{Z(P)}(\xi', \xi_n) \, d\xi' d\xi_n$$

$$= \int_{-\infty}^{\infty} \left(\int_{\mathbb{R}^{n-1}} \chi_{Z(P)}(\xi', \xi_n) \, d\xi' \right) d\xi_n = 0.$$

$\square$

Lemma 21.11. *Let $P(D) = \sum_{|\alpha| \leq m} a_\alpha D^\alpha$ be a linear partial differential operator with constant coefficients and of order m on $\mathbb{R}^n$. Then the operators $P(D) : H^{m,2} \to L^2(\mathbb{R}^n)$ and $P(D)^t : L^2(\mathbb{R}^n) \to H^{-m,2}$ are both injective.*

Proof Let $u \in H^{m,2}$ be such that

$$P(D)u = 0.$$

Then $u \in L^2(\mathbb{R}^n)$ by the Sobolev embedding theorem in Chapter 12. Taking the Fourier transform on both sides, we get

$$P(\xi)\hat{u}(\xi) = 0$$

for almost all $\xi \in \mathbb{R}^n$. By Lemma 21.10, $\hat{u} = 0$ a.e. on $\mathbb{R}^n$. Therefore $u = 0$. This proves that $P(D) : H^{m,2} \to L^2(\mathbb{R}^n)$ is injective. Now, let $u \in L^2(\mathbb{R}^n)$ be such that

$$P(D)^t u = 0.$$

Then, by Exercise 13.1, Exercise 13.2 and Plancherel's theorem,

$$(P(D)^t u, v) = (u, P(D)v) = (\hat{u}, P\hat{v}) = (\overline{P}\hat{u}, \hat{v}) = 0, \quad v \in H^{m,2}.$$

Therefore

$$(\overline{P}\hat{u}, \varphi) = 0, \quad \varphi \in \mathcal{S}.$$

Since $\mathcal{S}$ is dense in $L^2(\mathbb{R}^n)$, it follows that $\overline{P}\hat{u} = 0$. By Lemma 21.10, $\hat{u} = 0$ a.e. on $\mathbb{R}^n$. Therefore $u = 0$. So, $P(D)^t : L^2(\mathbb{R}^n) \to H^{-m,2}$ is injective. $\square$

Proof of Proposition 21.9 Suppose that $P(D) : H^{m,2} \to L^2(\mathbb{R}^n)$ is Fredholm with zero index. Then the range $R(P)$ of $P(D) : H^{m,2} \to L^2(\mathbb{R}^n)$ is closed in $L^2(\mathbb{R}^n)$. By Theorem 20.6, there exists a positive constant C such that

$$\|u\|_{m,2} \leq C\|P(D)u\|_2, \quad u \in H^{m,2}.$$

So,

$$0 \notin \overline{\{P(\xi) : \xi \in \mathbb{R}^n\}}.$$

For if this were not true, then there would be a sequence $\{\xi_k\}$ in $\mathbb{R}^n$ such that $P(\xi_k) \to 0$ as $k \to \infty$, and the analysis in Example 21.8 would apply to give a contradiction. Conversely, suppose that

$$0 \notin \overline{\{P(\xi) : \xi \in \mathbb{R}^n\}}.$$

Then there exists a positive constant C such that

$$|P(\xi)| \geq C, \quad \xi \in \mathbb{R}^n.$$

So, for all $u \in H^{m,2}$, by Plancherel's theorem,

$$\|P(D)u\|_2 = \|P\hat{u}\|_2 \geq C\|\hat{u}\|_2 = C\|u\|_2, \quad u \in H^{m,2}.$$

Therefore the range $R(P)$ is closed in $L^2(\mathbb{R}^n)$. The rest of the proof follows from Lemma 21.11. $\square$

The following result follows from Proposition 21.9.

Theorem 21.12. *Let $P(D)$ be a linear partial differential operator with constant coefficients and of order m on $\mathbb{R}^n$. Then, for $-\infty < s < \infty$, the operator $P(D) : H^{s,2} \to H^{s-m,2}$ is Fredholm with zero index if and only if*

$$0 \notin \overline{\{P(\xi) : \xi \in \mathbb{R}^n\}}.$$

Proof We only need to prove that $P(D) : H^{s,2} \to H^{s-m,2}$ is Fredholm with zero index if and only if $P(D) : H^{m,2} \to L^2(\mathbb{R}^n)$ is Fredholm with zero index. Suppose that $P(D) : H^{s,2} \to H^{s-m,2}$ is Fredholm with zero index. By Exercise 21.5, $P(D) = J_{m-s}P(D)J_{s-m} : H^{m,2} \to L^2(\mathbb{R}^n)$ is Fredholm with zero index. Similarly, the converse is true. $\square$

Remark 21.13. In fact, in Theorem 21.12, if $0 \notin \overline{\{P(\xi) : \xi \in \mathbb{R}^n\}}$, then $P(D) : H^{s,2} \to H^{s-m,2}$ is a bijection for $-\infty < s < \infty$.

Exercises

21.1. Give an example of a linear partial differential operator $P(D)$ on $\mathbb{R}^n$ such that $P(D)$ is elliptic, injective and there exists a sequence $\{\xi_k\}$ in $\mathbb{R}^n$ such that

$$P(\xi_k) \to 0$$

as $k \to \infty$.

21.2. Prove that a linear partial differential operator $P(D)$ with constant coefficients on $\mathbb{R}^n$ is elliptic if

$$0 \notin \overline{\{P(\xi) : \xi \in \mathbb{R}^n\}}.$$

Is the converse true?

21.3. Use the definition of S^0 to give another formula for $p_{\alpha,\beta}(\sigma)$ in Proposition 21.4.

21.4. Use the definition of S^0 and Peetre's inequality to fill in the details in deriving (21.6).

21.5. Let $P(D)$ be a linear partial differential operator with constant coefficients and of order m on $\mathbb{R}^n$. Prove that for all $s \in (-\infty, \infty)$, $P(D)$ and J_s commute in the sense that

$$P(D)J_s : H^{t,2} \to H^{s+t-m,2}$$

and

$$J_s P(D) : H^{t,2} \to H^{s+t-m,2}$$

are equal for all $t \in (-\infty, \infty)$.

Chapter 22

Symmetrically Global
Pseudo-Differential Operators

The pseudo-differential operators studied in this book are global operators on $\mathbb{R}^n$. The decay estimates on the associated symbols due to differentiations have hitherto been imposed only on the ξ-variable. We introduce in this chapter another class of pseudo-differential operators on $\mathbb{R}^n$ of which the symbols satisfy similar decay estimates due to differentiations with respect to the x-variable and the ξ-variable. They are appropriately called *symmetrically global* pseudo-differential operators. These operators are also known in the literature as SG operators with SG standing for *symbol-global*, *scattering operators* and *operators with exit at infinity*. For the sake of convenience and the close match with the title of this chapter, we simply call these operators SG operators and their symbols SG symbols.

Let $m_1, m_2 \in (-\infty, \infty)$. Then we let S^{m_1,m_2} be the set of all functions in $C^\infty(\mathbb{R}^n \times \mathbb{R}^n)$ such that for all multi-indices α and β, there exists a positive constant $C_{\alpha,\beta}$ for which

$$|(D_x^\alpha D_\xi^\beta \sigma)(x,\xi)| \leq C_{\alpha,\beta} \langle x \rangle^{m_2 - |\alpha|} \langle \xi \rangle^{m_1 - |\beta|}, \quad x, \xi \in \mathbb{R}^n.$$

A function in S^{m_1,m_2} is said to be a SG *symbol* of order (m_1, m_2). It is clear that if $\sigma \in S^{m_1,m_2}$ and $m_2 \leq 0$, then $\sigma \in S^{m_1}$, where S^{m_1} is the class of symbols of pseudo-differential operators studied before this chapter in the book. Let $\sigma \in S^{m_1,m_2}$. Then we define the SG *pseudo-differential operator* T_σ with symbol σ by

$$(T_\sigma \varphi)(x) = (2\pi)^{-n/2} \int_{\mathbb{R}^n} e^{ix \cdot \xi} \sigma(x,\xi) \hat{\varphi}(\xi)\, d\xi, \quad x \in \mathbb{R}^n, \tag{22.1}$$

for all functions φ in the Schwartz space $\mathcal{S}$. It can be proved easily that $T_\sigma : \mathcal{S} \to \mathcal{S}$ is a continuous linear mapping. (See Propositions 6.7 and 11.1.)

We begin with the product formula.

Theorem 22.1. *Let $\sigma \in S^{m_1,m_2}$ and $\tau \in S^{\mu_1,\mu_2}$. Then $T_\sigma T_\tau = T_\lambda$, where $\lambda \in S^{m_1+\mu_1,m_2+\mu_2}$ and*

$$\lambda \sim \sum_\mu \frac{(-i)^{|\mu|}}{\mu!}(\partial_\xi^\mu \sigma)(\partial_x^\mu \tau).$$

Here the asymptotic expansion means that for every positive integer M, there exists a positive integer N such that

$$\lambda - \sum_{|\mu|<N} \frac{(-i)^{|\mu|}}{\mu!}(\partial_\xi^\mu \sigma)(\partial_x^\mu \tau) \in S^{m_1+\mu_1-M,m_2+\mu_2-M}.$$

The next important result is the formula for the formal adjoint.

Theorem 22.2. *Let $\sigma \in S^{m_1,m_2}$. Then the formal adjoint T_σ^* of T_σ is a SG pseudo-differential operator T_τ, where $\tau \in S^{m_1,m_2}$ and*

$$\tau \sim \sum_\mu \frac{(-i)^{|\mu|}}{\mu!}\partial_x^\mu \partial_\xi^\mu \overline{\sigma}.$$

Here the asymptotic expansion means that for every positive integer M, there exists a positive integer N such that

$$\tau - \sum_{|\mu|<N} \frac{(-i)^{|\mu|}}{\mu!}\partial_x^\mu \partial_\xi^\mu \overline{\sigma} \in S^{m_1-M,m_2-M}.$$

As in Chapter 11, we can extend the definition of a SG pseudo-differential operator from the Schwartz space $\mathcal{S}$ to the space $\mathcal{S}'$ of all tempered distributions. Let $\sigma \in S^{m_1,m_2}$. Then for all u in $\mathcal{S}'$, we define the linear functional $T_\sigma : \mathcal{S} \to \mathbb{C}$ by

$$(T_\sigma u)(\varphi) = u(\overline{T_\sigma^* \overline{\varphi}}), \quad \varphi \in \mathcal{S}.$$

As an analog of Proposition 11.6, T_σ maps $\mathcal{S}'$ into $\mathcal{S}'$ continuously. The more interesting result is the following L^p-boundedness theorem. It follows from the fact that each symbol in $S^{0,0}$ is in S^0 and Theorem 11.7.

Theorem 22.3. *Let $\sigma \in S^{0,0}$. Then $T_\sigma : L^p(\mathbb{R}^n) \to L^p(\mathbb{R}^n)$ is a bounded linear operator for $1 < p < \infty$.*

We now come to the ellipticity of SG pseudo-differential operators. Let $\sigma \in S^{m_1,m_2}$, $-\infty < m_1, m_2 < \infty$. Then σ is said to be SG *elliptic* if there exist positive constants C and R such that

$$|\sigma(x,\xi)| \geq C\langle x \rangle^{m_2} \langle \xi \rangle^{m_1}, \quad |x|^2 + |\xi|^2 \geq R^2.$$

The following theorem tells us that SG elliptic operators have SG parametrices.

Theorem 22.4. *Let* $\sigma \in S^{m_1,m_2}$, $-\infty < m_1, m_2 < \infty$, *be SG elliptic. Then there exists a symbol* τ *in* $S^{-m_1,-m_2}$ *such that*

$$T_\tau T_\sigma = I + R$$

and

$$T_\sigma T_\tau = I + S,$$

where R *and* S *are infinitely smoothing in the sense that they are SG pseudo-differential operators with symbols in* $\bigcap_{k_1,k_2 \in \mathbb{R}} S^{k_1,k_2}$.

The SG pseudo-differential operator T_τ in the preceding theorem is known as a SG *parametrix* of T_σ.

The next development of the theory is to determine the domain of a SG elliptic pseudo-differential operator. We need Sobolev spaces for SG pseudo-differential operators for this task. For $s_1, s_2 \in (-\infty, \infty)$, we let J_{s_1,s_2} be the *Bessel potential* of order (s_1, s_2) defined by

$$J_{s_1,s_2} = T_{\sigma_{s_1,s_2}},$$

where

$$\sigma_{s_1,s_2}(x,\xi) = \langle x \rangle^{-s_2} \langle \xi \rangle^{-s_1}, \quad x, \xi \in \mathbb{R}^n.$$

Obviously, $\sigma_{s_1,s_2} \in S^{-s_1,-s_2}$. It is left as Exercise 22.6 to show that the mapping $J_{s_1,s_2} : S' \to S'$ is a bijection for $-\infty < s_1, s_2 < \infty$, and

$$J_{s_1,s_2}^{-1} = J_{-s_1,0} J_{0,-s_2} \tag{22.2}$$

and hence, by Theorem 22.1, J_{s_1,s_2}^{-1} is a SG pseudo-differential operator of order $(-s_1, -s_2)$.

For $1 < p < \infty$, and $-\infty < s_1, s_2 < \infty$, the L^p-*Sobolev space* $H^{s_1,s_2,p}$ of order (s_1, s_2) is defined by

$$H^{s_1,s_2,p} = \{ u \in S' : J_{-s_1,-s_2} u \in L^p(\mathbb{R}^n) \}.$$

Then $H^{s_1,s_2,p}$ is a Banach space in which the norm $\| \ \|_{s_1,s_2,p}$ is given by

$$\|u\|_{s_1,s_2,p} = \|J_{-s_1,-s_2} u\|_p, \quad u \in H^{s_1,s_2,p}.$$

Obviously,

$$H^{0,0,p} = L^p(\mathbb{R}^n).$$

Theorem 22.3 can now be improved to the following result. Its proof is left as an exercise.

Theorem 22.5. *Let* $\sigma \in S^{m_1,m_2}$, $-\infty < m_1, m_2 < \infty$, *be a SG symbol. Then* $T_\sigma : H^{s_1,s_2,p} \to H^{s_1-m_1,s_2-m_2,p}$ *is a bounded linear operator for* $1 < p < \infty$ *and* $-\infty < s_1, s_2 < \infty$.

We have the following Sobolev embedding theorem for SG pseudo-differential operators. (See Theorem 12.5 and Remark 12.10.) In Theorem 12.5, we have a bounded linear operator of which the norm is at most one. There is a short proof if we relax the requirement of the norm. The proof of the following theorem is also not difficult if we just need the inclusion to be a bounded linear operator. This is a good exercise.

Theorem 22.6. *Let $s_1, s_2, t_1, t_2 \in (-\infty, \infty)$ be such that $s_1 \leq t_1$ and $s_2 \leq t_2$. Then $H^{t_1, t_2, p} \subseteq H^{s_1, s_2, p}$ and the inclusion $i : H^{t_1, t_2, p} \hookrightarrow H^{s_1, s_2, p}$ is a bounded linear operator.*

The most important feature of the Sobolev spaces for SG pseudo-differential operators is the following compact embedding theorem, which is not true for the Sobolev spaces in Chapter 12.

Theorem 22.7. *Let $s_1, s_2, t_1, t_2 \in (-\infty, \infty)$ be such that $s_1 < t_1$ and $s_2 < t_2$. Then the inclusion $i : H^{t_1, t_2, p} \hookrightarrow H^{s_1, s_2, p}$ is a compact operator.*

A proof of Theorem 22.7 depends on the following result in [Wong (1983, 1994)]. It is a result in [Grushin (1970)] at least for the L^2-case and we skip the proof.

Theorem 22.8. *Let $\sigma \in S^m$, $m \in (-\infty, \infty)$, be such that*

$$\lim_{|x| \to \infty} C_{\alpha, \beta}(x) = 0$$

for all multi-indices α and β. Then for every positive number ε, the operator $T_\sigma : H^{s+m, p} \to H^{s-\varepsilon, p}$ is compact for $-\infty < s < \infty$ and $1 < p < \infty$.

The following corollary follows immediately. We leave it as an exercise.

Corollary 22.9. *For every positive number ε, $J_{\varepsilon, \varepsilon} : L^p(\mathbb{R}^n) \to L^p(\mathbb{R}^n)$ is a compact operator for $1 < p < \infty$.*

Proof of Theorem 22.7 Let ε be a positive number such that

$$t_1 - s_1 - \varepsilon > 0$$

and

$$t_2 - s_2 - \varepsilon > 0.$$

Then, by Theorem 22.1 and (22.2), $J_{\varepsilon, \varepsilon}^{-1} J_{-s_1, -s_2}$ is a SG pseudo-differential operator of order $(s_1 + \varepsilon, s_2 + \varepsilon)$. So,

$$J_{\varepsilon, \varepsilon}^{-1} J_{-s_1, -s_2} : H^{t_1, t_2, p} \to H^{t_1 - s_1 - \varepsilon, t_2 - s_2 - \varepsilon}$$

is a bounded linear operator, the inclusion

$$i : H^{t_1 - s_1 - \varepsilon, t_2 - s_2 - \varepsilon, p} \hookrightarrow L^p(\mathbb{R}^n)$$

is a bounded linear operator, the operator

$$J_{\varepsilon, \varepsilon} : L^p(\mathbb{R}^n) \to L^p(\mathbb{R}^n)$$

is compact, and the operator

$$J_{s_1, s_2} : L^p(\mathbb{R}^n) \to H^{s_1, s_2, p}$$

is a bounded linear operator. Thus,

$$i = J_{s_1, s_2} J_{\varepsilon, \varepsilon} i J_{\varepsilon, \varepsilon}^{-1} J_{-s_1, -s_2} : H^{t_1, t_2, p} \to H^{s_1, s_2, p}$$

is a compact operator. □

The first main result in this chapter is the following theorem.

Theorem 22.10. *Let* $\sigma \in S^{m_1, m_2}$, $-\infty < m_1, m_2 < \infty$, *be SG elliptic. Then for all* $s_1, s_2 \in (-\infty, \infty)$, $T_\sigma : H^{s_1, s_2, p} \to H^{s_1 - m_1, s_2 - m_2, p}$ *is a Fredholm operator for* $1 < p < \infty$.

Proof Since σ is SG elliptic, it follows from Theorem 22.4 that there exists a symbol τ in $S^{-m_1, -m_2}$ such that

$$T_\tau T_\sigma = I + R$$

and

$$T_\sigma T_\tau = I + S,$$

where R and S are infinitely smoothing in the sense that they are SG pseudo-differential operators with symbols in $\bigcap_{k_1, k_2 \in \mathbb{R}} S^{k_1, k_2}$. So, for all positive numbers t_1 and t_2, the linear operator $R : L^p(\mathbb{R}^n) \to L^p(\mathbb{R}^n)$ is the same as the composition of the linear operators $R : L^p(\mathbb{R}^n) \to H^{t_1, t_2, p}$ and $i : H^{t_1, t_2, p} \hookrightarrow L^p(\mathbb{R}^n)$. Since $R : L^p(\mathbb{R}^n) \to H^{t_1, t_2, p}$ is bounded by Theorem 22.5 and $i : H^{t_1, t_2, p} \hookrightarrow L^p(\mathbb{R}^n)$ is compact by Theorem 22.7, it follows that $R : L^p(\mathbb{R}^n) \to L^p(\mathbb{R}^n)$ is compact. Similarly, $S : L^p(\mathbb{R}^n) \to L^p(\mathbb{R}^n)$ is compact. So, by Theorem 20.5 of Atkinson, $T_{\sigma,0}$ is Fredholm. □

We give an analog of Theorem 21.7 for SG pseudo-differential operators.

Theorem 22.11. *Let* $\sigma \in S^{m_1, m_2}$, $m_1, m_2 \in (-\infty, \infty)$, *be such that the operator* $T_\sigma : H^{s_1, s_2, p} \to H^{s_1 - m_1, s_2 - m_2, p}$ *is Fredholm for* $s_1, s_2 \in (-\infty, \infty)$ *and* $1 < p < \infty$. *Then* σ *is SG elliptic.*

We first establish an analog of Theorem 21.6 for SG pseudo-differential operators.

Theorem 22.12. *Let $\sigma \in S^{0,0}$ be such that $T_\sigma : L^p(\mathbb{R}^n) \to L^p(\mathbb{R}^n)$ is Fredholm for $1 < p < \infty$. Then we can find positive constants C and R such that*

$$|\sigma(x, \xi)| \geq C$$

for all x and ξ in $\mathbb{R}^n$ with

$$|x|^2 + |\xi|^2 \geq R^2.$$

Proof By Atkinson's theorem, i.e., Theorem 20.5, we can find a bounded linear operator S on $L^p(\mathbb{R}^n)$ and a compact operator K on $L^p(\mathbb{R}^n)$ such that

$$ST_\sigma = I + K.$$

Let M be the subset of $\mathbb{R}^n \times \mathbb{R}^n$ given by

$$\left\{ (x, \xi) \in \mathbb{R}^n : |\sigma(x, \xi)| \leq \frac{1}{2\|S\|} \right\}.$$

We first suppose that M is bounded, i.e., there exists a positive number R such that

$$|(x, \xi)| \leq R, \quad (x, \xi) \in M.$$

So,

$$x^2 + \xi^2 \geq R^2 \Rightarrow (x, \xi) \notin M \Rightarrow |\sigma(x, \xi)| \geq \frac{1}{2\|S\|}.$$

Thus, the theorem is proved. If M is unbounded, then there exists a sequence $\{(x_k, \xi_k)\}$ in M such that

$$|(x_k, \xi_k)| \to \infty$$

as $k \to \infty$ and

$$|\sigma(x_k, \xi_k)| \leq \frac{1}{2\|S\|}, \quad k = 1, 2, \ldots.$$

So, there exists a subsequence of $\{(x_k, \xi_k)\}$, again denoted by $\{(x_k, \xi_k)\}$, such that

$$|\sigma(x_k, \xi_k)| \to \sigma_\infty$$

as $k \to \infty$, where σ_∞ is some complex number. Thus,

$$|\sigma_\infty| \leq \frac{1}{2\|S\|}.$$

Now, for $k = 1, 2, \ldots$, let $\lambda_k = |(x_k, \xi_k)|$. Since (21.3) is equally valid for $S^{0,0}$, it follows that

$$R_{\lambda_k, \tau} \left(x_k, \frac{\xi_k}{|\xi_k|} \right)^{-1} T_\sigma R_{\lambda_k, \tau} \left(x_k, \frac{\xi_k}{|\xi_k|} \right) = T_{\sigma_k, \tau},$$

where

$$\sigma_{\sigma_k, \tau}(x, \eta) = \sigma \left(x_k + \lambda_k^{-\tau} x, \lambda_k \frac{\xi_k}{|\xi_k|} + \lambda_k^\tau \eta \right), \quad x, \eta \in \mathbb{R}^n.$$

By Exercise 22.13, (21.5) is valid also for $S^{0,0}$ if we replace $p_{\alpha,\beta}$ by the corresponding norm $q_{\alpha,\beta}$ in $S^{0,0}$. So, for all multi-indices α and β, there exists a positive constant C_β such that

$$|(\partial_x^\alpha \partial_\eta^\beta \sigma_{\lambda_k,\tau})(x, \eta)| \leq C_\beta q_{\alpha,\beta} \langle \eta \rangle^{|\beta|} \lambda_k^{-\tau|\alpha|} \lambda_k^{-(1-2\tau)|\beta|},$$

and the rest of the proof is exactly the same as that of Theorem 21.6. $\square$

Using Theorem 22.12, the proof of Theorem 22.11 is similar to that of Theorem 21.7 and is best left as an exercise.

We end this chapter with an index formula for SG pseudo-differential operators. We need a lemma.

Lemma 22.13. $\cap_{k_1, k_2 \in \mathbb{R}} S^{k_1, k_2} = S$.

Proof It is easy to check that $S \subseteq \cap_{k_1, k_2 \in \mathbb{R}} S^{k_1, k_2}$ and is left as an exercise. Let $\sigma \in \cap_{k_1, k_2 \in \mathbb{R}} S^{k_1, k_2}$. Then for all multi-indices α, β, γ and δ, let k_1 and k_2 be real numbers such that

$$|\alpha| + k_2 - |\gamma| \leq 0$$

and

$$|\beta| + k_1 - |\delta| \leq 0.$$

Since $\sigma \in S^{k_1, k_2}$, there exists a positive constant $C_{k_1, k_2, \gamma, \delta}$ such that

$$\sup_{x, \xi \in \mathbb{R}^n} |x^\alpha \xi^\beta (D_x^\gamma D_\xi^\delta \sigma)(x, \xi)|$$

$$\leq \sup_{x, \xi \in \mathbb{R}^n} (|x|^{|\alpha|} |\xi|^{|\beta|} C_{k_1, k_2, \gamma, \delta} \langle x \rangle^{k_2 - |\gamma|} \langle \xi \rangle^{k_1 - |\delta|})$$

$$\leq C_{k_1, k_2, \gamma, \delta} \sup_{x, \xi \in \mathbb{R}^n} (\langle x \rangle^{|\alpha| + k_2 - \gamma} \langle \xi \rangle^{|\beta| + k_1 - |\delta|}) < \infty.$$

Therefore $\sigma \in S$. $\square$

Let $\sigma \in S^{0,0}$ be a SG elliptic symbol. Let $\tau \in S^{0,0}$ be such that

$$T_\tau T_\sigma = I - T_1$$

and

$$T_\sigma T_\tau = I - T_2,$$

where T_j is a SG pseudo-differential operator with symbol τ_j in $\cap_{k_1,k_2 \in \mathbb{R}} S^{k_1,k_2}$, $j = 1, 2$. Therefore for $j = 1, 2$, $\tau_j \in \mathcal{S}$ by Lemma 22.13. Let $\varphi \in \mathcal{S}$. Then for $j = 1, 2$,

$$(T_j \varphi)(x) = (2\pi)^{-n/2} \int_{\mathbb{R}^n} e^{ix \cdot \xi} \tau_j(x, \xi) \hat{\varphi}(\xi) \, d\xi$$

$$= (2\pi)^{-n/2} \int_{\mathbb{R}^n} T_{-x}(\mathcal{F}_2 \tau_j)(x, y) \, \varphi(y) \, dy$$

for all $x \in \mathbb{R}^n$, where $\mathcal{F}_2$ denotes the Fourier transform with respect to the *second* variable. So, for $j = 1, 2$,

$$(T_j \varphi)(x) = (2\pi)^{-n/2} \int_{\mathbb{R}^n} (\mathcal{F}_2 \tau_j)(x, y - x) \, \varphi(y) \, dy$$

$$= (2\pi)^{-n/2} \int_{\mathbb{R}^n} (\mathcal{F}_2^{-1} \tau_j)(x, x - y) \, \varphi(y) \, dy$$

for all $x \in \mathbb{R}^n$. For $j = 1, 2$, $(\mathcal{F}_2^{-1} \tau_j)(x, x - y)$ is a Schwartz function of (x, y) in $\mathbb{R}^n \times \mathbb{R}^n$. So, by Theorem 20.12, $T_j : L^2(\mathbb{R}^n) \to L^2(\mathbb{R}^n)$ is a trace class operator with

$$\text{tr}(T_j) = (2\pi)^{-n/2} \int_{\mathbb{R}^n} (\mathcal{F}_2^{-1} \tau_j)(x, 0) \, dx$$

$$= (2\pi)^{-n} \int_{\mathbb{R}^n} \int_{\mathbb{R}^n} e^{i0 \cdot \xi} \tau_j(x, \xi) \, dx \, d\xi$$

$$= (2\pi)^{-n} \int_{\mathbb{R}^n} \int_{\mathbb{R}^n} \tau_j(x, \xi) \, dx \, d\xi,$$

for $j = 1, 2$. By Theorem 20.13,

$$i(T_\sigma) = \text{tr}(T_1) - \text{tr}(T_2) = (2\pi)^{-n} \int_{\mathbb{R}^n} \int_{\mathbb{R}^n} (\tau_1(x, \xi) - \tau_2(x, \xi)) \, dx \, d\xi.$$

Finally, let $\sigma \in S^{m_1, m_2}$ be such that the corresponding SG pseudo-differential operator $T_\sigma : H^{s_1, s_2, 2} \to H^{s_1 - m_1, s_2 - m_2, 2}$ is a Fredholm operator for $-\infty < m_1, m_2 < \infty$ and $-\infty < s_1, s_2 < \infty$. Then, by Theorem 22.11, σ is elliptic. So, there exists a symbol $\tau \in S^{-m_1, -m_2}$ such that

$$T_\tau T_\sigma = I - T_1' \tag{22.3}$$

and

$$T_\sigma T_\tau = I - T_2', \tag{22.4}$$

where T_1' and T_2' are SG pseudo-differential operators with, respectively, symbols τ_1' and τ_2' in $\cap_{k_1,k_2 \in \mathbb{R}} S^{k_1,k_2}$. Now, the operator $J_{m_1-s_1,m_2-s_2} T_\sigma J_{s_1,s_2}$ is a SG pseudo-differential operator with symbol in $S^{0,0}$. By the analysis carried out in the preceding paragraph, Theorem 20.7 and the fact that J_{s_1,s_2} is a surjective isometry for all s_1 and s_2 in $\mathbb{R}$, $J_{m_1-s_1,m_2-s_2} T_\sigma J_{s_1,s_2} : L^2(\mathbb{R}^n) \to L^2(\mathbb{R}^n)$ is a Fredholm operator such that

$$i(J_{m_1-s_1,m_2-s_2} T_\sigma J_{s_1,s_2}) = i(T_\sigma). \tag{22.5}$$

By (22.3),

$$(J_{-s_1,-s_2} T_\tau J_{s_1-m_1,s_2-m_2})(J_{m_1-s_1,m_2-s_2} T_\sigma J_{s_1,s_2})$$
$$= J_{-s_1,-s_2} T_\tau T_\sigma J_{s_1,s_2}$$
$$= J_{-s_1,-s_2}(I - T_1') J_{s_1,s_2}$$
$$= I - J_{-s_1,-s_2} T_1' J_{s_1,s_2}.$$

Moreover, by (22.4),

$$(J_{m_1-s_1,m_2-s_2} T_\sigma J_{s_1,s_2})(J_{-s_1,-s_2} T_\tau J_{s_1-m_1,s_2-m_2})$$
$$= J_{m_1-s_1,m_2-s_2} T_\sigma T_\tau J_{s_1-m_1,s_2-m_2}$$
$$= J_{m_1-s_1,m_2-s_2}(I - T_2') J_{s_1-m_1,s_2-m_2}$$
$$= I - J_{m_1-s_1,m_2-s_2} T_2' J_{s_1-m_1,s_2-m_2}.$$

By Theorem 20.12, the equation (22.5) and Theorem 20.7, we get

$$i(T_\sigma) = \text{tr}(J_{-s_1,-s_2} T_1' J_{s_1,s_2}) - \text{tr}(J_{m_1-s_1,m_2-s_2} T_2' J_{s_1-m_1,s_2-m_2})$$
$$= \text{tr}(T_1') - \text{tr}(T_2').$$

Using the same analysis as in the preceding paragraph, we conclude that

$$i(T_\sigma) = (2\pi)^{-n} \int_{\mathbb{R}^n} \int_{\mathbb{R}^n} (\tau_1'(x,\xi) - \tau_2'(x,\xi))\, dx\, d\xi.$$

Exercises

22.1. Prove that every linear partial differential operator with constant coefficients on $\mathbb{R}^n$ is a SG pseudo-differential operator.

22.2 What is the SG order of the linear partial differential operator $\sum_{|\alpha| \leq m} a_\alpha D^\alpha$ with constant coefficients on $\mathbb{R}^n$?

22.3. Prove that every linear partial differential operator $P(D)$ on $\mathbb{R}^n$ is SG elliptic if and only if

$$0 \notin \overline{\{P(\xi) : \xi \in \mathbb{R}^n\}}.$$

22.4. Let $\sigma \in S^{m_1,m_2}$, $-\infty < m_1, m_2 < \infty$. Prove that $T_\sigma : \mathcal{S} \to \mathcal{S}$ is a continuous linear mapping.

22.5. For $-\infty < m_1, m_2 < \infty$, prove that if $\sigma \in S^{m_1,m_2}$ and $m_2 \leq 0$, then $\sigma \in S^{m_1}$.

22.6. Prove that for $-\infty < s_1, s_2 < \infty$, the mapping $J_{s_1,s_2} : \mathcal{S}' \to \mathcal{S}'$ is a bijection and

$$J_{s_1,s_2}^{-1} = J_{-s_1,0} J_{0,-s_2}.$$

22.7 Prove that for $1 < p < \infty$ and $-\infty < s < \infty$, $H^{s,0,p} = H^{s,p}$.

22.8. Prove that for $1 < p < \infty$ and $-\infty < s_1, s_2 < \infty$, the mapping $J_{-s_1,-s_2} : H^{s_1,s_2,p} \to L^p(\mathbb{R}^n)$ is a surjective isometry.

22.9. Prove Theorem 22.5.

22.10. Let $s_1, s_2, t_1, t_2 \in (-\infty, \infty)$ be such that $s_1 \leq t_1$ and $s_2 \leq t_2$. Prove that $H^{t_1,t_2,p} \subseteq H^{s_1,s_2,p}$ and the inclusion $i : H^{t_1,t_2,p} \hookrightarrow H^{s_1,s_2,p}$ is a bounded linear operator.

22.11. Prove Corollary 22.9.

22.12. Prove that, for $-\infty < s_1, s_2 < \infty$ and $1 < p < \infty$, the Schwartz space $\mathcal{S}$ is dense in $H^{s_1,s_2,p}$.

22.13. Prove (21.5) for symbols in $S^{0,0}$, where $p_{\alpha,\beta}$ is now replaced by the corresponding norm $q_{\alpha,\beta}$ in $S^{0,0}$.

22.14. Prove that if T_σ and T_τ are SG elliptic pseudo-differential operators, then $T_\sigma T_\tau$ is a SG pseudo-differential operator.

22.15. Prove that if T_σ is a SG elliptic pseudo-differential operator, then T_σ^* is also a SG elliptic pseudo-differential operator.

22.16. Prove Theorem 22.11.

22.17. Prove that $\mathcal{S} \subseteq \cap_{k_1,k_2 \in \mathbb{R}} S^{k_1,k_2}$.

22.18. Let $A : L^2(\mathbb{R}^n) \to L^2(\mathbb{R}^n)$ be an operator of finite rank such that the range $R(A)$ of A is contained in $\mathcal{S}$. Prove that A is a SG pseudo-differential operator with symbol in $\cap_{k_1,k_2 \in \mathbb{R}} S^{k_1,k_2}$.

Chapter 23

Spectral Invariance of Symmetrically Global Pseudo-Differential Operators

This chapter contains an application of the equivalence of the ellipticity and Fredholmness of SG pseudo-differential operators on $L^p(\mathbb{R}^n)$, $1 < p < \infty$. To motivate the topic, let us first observe that, in view of Theorems 22.2 and 22.5, the set of SG pseudo-differential operators with symbols in $S^{0,0}$ is an algebra of bounded linear operators on $L^p(\mathbb{R}^n)$ for $1 < p < \infty$. In the case when $p = 2$, the algebra is in fact a $*$-algebra. Suppose that $\sigma \in S^{0,0}$ is such that the corresponding SG pseudo-differential operator $T_\sigma : L^p(\mathbb{R}^n) \to L^p(\mathbb{R}^n)$ is bijective, where $1 < p < \infty$. The problem is to determine whether or not the inverse is also a SG pseudo-differential operator with symbol in $S^{0,0}$. This is known as the *spectral invariance* problem. See Exercise 23.4 for an explanation of the terminology.

We begin with the L^2-case.

Theorem 23.1. *Let $\sigma \in S^{0,0}$ be such that the corresponding SG pseudo-differential operator $T_\sigma : L^2(\mathbb{R}^n) \to L^2(\mathbb{R}^n)$ is bijective. Then its inverse $T_\sigma^{-1} : L^2(\mathbb{R}^n) \to L^2(\mathbb{R}^n)$ is also a SG pseudo-differential operator with symbol in $S^{0,0}$.*

Proof By bijectivity, $T_\sigma : L^2(\mathbb{R}^n) \to L^2(\mathbb{R}^n)$ is Fredholm and

$$i(T_\sigma) = 0.$$

By Theorem 22.12, T_σ is SG elliptic. By Theorem 22.4, there exists a symbol τ in $S^{0,0}$ such that

$$T_\tau T_\sigma = I + R$$

and

$$T_\sigma T_\tau = I + S,$$

where R and S are infinitely smoothing in the sense that they are SG pseudo-differential operators with symbols in $\cap_{k_1,k_2 \in \mathbb{R}} S^{k_1,k_2}$. By Atkinson's

177

Theorem 20.5, T_τ is Fredholm and hence by Theorem 22.4, T_τ is elliptic. Also, by Remark 20.4 and Theorem 20.7,

$$i(T_\tau) + i(T_\sigma) = i(T_\tau T_\sigma) = i(I + R) = 0.$$

So,

$$i(T_\tau) = 0.$$

Now, let $u \in N(T_\tau)$. Then

$$T_\tau u = 0 \Rightarrow T_\sigma T_\tau u = 0 \Rightarrow (I + S)u = 0 \Rightarrow u = -Su.$$

Since $\cap_{s_1, s_2 \in \mathbb{R}} H^{s_1, s_2, 2} = S$, it follows that

$$u = -Su \in S.$$

Similarly, the null space $N(T_\tau^t)$ of the true adjoint $T_\tau^t : L^2(\mathbb{R}^n) \to L^2(\mathbb{R}^n)$ of $T_\tau : L^2(\mathbb{R}^n) \to L^2(\mathbb{R}^n)$ is also a subspace of S. Now, we write

$$L^2(\mathbb{R}^n) = N(T_\tau) \oplus N(T_\tau)^\perp$$

and

$$L^2(\mathbb{R}^n) = N(T_\tau^t) \oplus R_{L^2}(T_\tau),$$

where $R_{L^2}(T_\tau)$ is the range of $T_\tau : L^2(\mathbb{R}^n) \to L^2(\mathbb{R}^n)$. Let $P = iF\pi$, where π is the projection of $L^2(\mathbb{R}^n)$ onto $N(T_\tau)$, F is an isomorphism of $N(T_\tau)$ onto $N(T_\tau^t)$ and i is the inclusion of $N(T_\tau^t)$ into $L^2(\mathbb{R}^n)$. Then the operator $T_\tau + P : L^2(\mathbb{R}^n) \to L^2(\mathbb{R}^n)$ is a bijective parametrix of T_σ. Therefore, without loss of generality, we may assume that the parametrix $T_\tau : L^2(\mathbb{R}^n) \to L^2(\mathbb{R}^n)$ is bijective. So, $I + R$ is bijective. In fact,

$$(I + R)^{-1} = I + K,$$

where K maps $L^2(\mathbb{R}^n)$ into S. Indeed, there exists a bounded linear operator $K : L^2(\mathbb{R}^n) \to L^2(\mathbb{R}^n)$ such that

$$(I + R)(I + K) = I.$$

So,

$$K = -R - RK$$

and

$$K^* = -R^* - K^* R^*.$$

It is obvious that K and K^* map $L^2(\mathbb{R}^n)$ into S. Thus, by Exercise 23.2, the kernel of $K : L^2(\mathbb{R}^n) \to L^2(\mathbb{R}^n)$ is a Schwartz function on $\mathbb{R}^n \times \mathbb{R}^n$, and hence, by Exercise 23.3, K is a SG pseudo-differential operator with symbol in $\cap_{k_1,k_2 \in \mathbb{R}} S^{k_1,k_2}$. Thus,

$$T_\sigma^{-1} T_\tau^{-1} = I + K$$

or equivalently

$$T_\sigma^{-1} = (I + K) T_\tau$$

and this completes the proof. $\qquad\square$

We can now give the L^p-version of the spectral invariance of SG pseudo-differential operators.

Theorem 23.2. *Let $\sigma \in S^{0,0}$ be such that the pseudo-differential operator $T_\sigma : L^p(\mathbb{R}^n) \to L^p(\mathbb{R}^n)$ is invertible, where $1 < p < \infty$. Then its inverse $T_\sigma^{-1} : L^p(\mathbb{R}^n) \to L^p(\mathbb{R}^n)$ is also a pseudo-differential operator with symbol in $S^{0,0}$.*

Proof Since $T_\sigma : L^p(\mathbb{R}^n) \to L^p(\mathbb{R}^n)$ is bijective, it follows that it is Fredholm. So, by Theorem 22.4, σ is an elliptic symbol in $S^{0,0}$. Thus, by Theorem 22.4, $T_\sigma : L^2(\mathbb{R}^n) \to L^2(\mathbb{R}^n)$ is Fredholm. So, there exists a symbol τ in $S^{0,0}$ such that

$$T_\tau T_\sigma = I + R$$

and

$$T_\sigma T_\tau = I + S,$$

where R and S are infinitely smoothing. Let $u \in L^2(\mathbb{R}^n)$ be such that $T_\sigma u = 0$. Then

$$T_\tau T_\sigma u = 0 \Rightarrow (I + R)u = 0 \Rightarrow u = -Ru \in S. \qquad (23.1)$$

So, u is also in the null space of $T_\sigma : L^p(\mathbb{R}^n) \to L^p(\mathbb{R}^n)$. Since $T_\sigma : L^p(\mathbb{R}^n) \to L^p(\mathbb{R}^n)$ is injective, it follows that $u = 0$. Therefore $T_\sigma : L^2(\mathbb{R}^n) \to L^2(\mathbb{R}^n)$ is injective. To show that $T_\sigma : L^2(\mathbb{R}^n) \to L^2(\mathbb{R}^n)$ is surjective, let u be a function in the null space $N(T_\sigma^t)$ of the adjoint $T_\sigma^t : L^2(\mathbb{R}^n) \to L^2(\mathbb{R}^n)$ of $T_\sigma : L^2(\mathbb{R}^n) \to L^2(\mathbb{R}^n)$. Then

$$T_\sigma^t u = 0 \Rightarrow T_{\sigma^*} u = 0,$$

where σ^* in $S^{0,0}$ is the symbol of the formal adjoint of T_σ. Since σ^* is elliptic, we can use a left parametrix of T_{σ^*} as in (23.1) to conclude that $u = 0$. This proves that $T_\sigma : L^2(\mathbb{R}^n) \to L^2(\mathbb{R}^n)$ is surjective and hence bijective. So, by Theorem 23.1, the inverse $T_\sigma^{-1} : L^2(\mathbb{R}^n) \to L^2(\mathbb{R}^n)$ is a SG pseudo-differential operator T_τ with symbol τ in $S^{0,0}$. Since

$$T_\sigma^{-1} \varphi = T_\tau \varphi, \qquad \varphi \in S,$$

and S is dense in $L^p(\mathbb{R}^n)$, it follows that $T_\sigma^{-1} = T_\tau$ on $L^p(\mathbb{R}^n)$. $\qquad\square$

Exercises

23.1. Prove that $\bigcap_{s_1,s_2\in\mathbb{R}} H^{s_1,s_2,2} = S$. (Compare this exercise with Exercise 12.13.)

23.2. Prove that if $A : L^2(\mathbb{R}^n) \to L^2(\mathbb{R}^n)$ and $A^t : L^2(\mathbb{R}^n) \to L^2(\mathbb{R}^n)$ are bounded linear operators such that A and A^t map $L^2(\mathbb{R}^n)$ into S, then the kernel of A is a Schwartz function on $\mathbb{R}^n \times \mathbb{R}^n$. (Hint: Use Exercise 20.6.)

23.3. Let $A : L^2(\mathbb{R}^n) \to L^2(\mathbb{R}^n)$ be a Hilbert–Schmidt operator such that its kernel is a Schwartz function on $\mathbb{R}^n \times \mathbb{R}^n$. Prove that A is a SG pseudo-differential operator with symbol in $\bigcap_{k_1,k_2\in\mathbb{R}} S^{k_1,k_2}$.

23.4. Let $B(X)$ be the Banach algebra of all bounded linear operators on a complex Banach space X. Let $\mathcal{A}$ be a sub-algebra of $B(X)$ containing the identity operator I on X. Then for every element A in $\mathcal{A}$, we define the *resolvent set* $\rho_{\mathcal{A}}(A)$ of A with respect to $\mathcal{A}$ to be the set of all complex numbers λ such that $A - \lambda I$ has an inverse, which also lies in $\mathcal{A}$. The *spectrum* $\Sigma_{\mathcal{A}}(A)$ of A with respect to $\mathcal{A}$ is defined to be the complement in $\mathbb{C}$ of the resolvent set $\rho_{\mathcal{A}}(A)$.

(i) Prove that

$$\Sigma_{B(X)}(A) \subseteq \Sigma_{\mathcal{A}}(A), \quad A \in \mathcal{A}.$$

(ii) A sub-algebra $\mathcal{A}$ of $B(X)$ containing the identity operator I on X is said to be nontrivial if $\mathcal{A} \neq B(X)$. Give an example to show that there exist complex Banach spaces X and nontrivial sub-algebras $\mathcal{A}$ of $B(X)$ for which

$$\Sigma_{B(X)}(A) = \Sigma_{\mathcal{A}}(A), \quad A \in \mathcal{A}. \tag{23.2}$$

(A nontrivial sub-algebra $\mathcal{A}$ of $B(X)$ containing the identity operator I on X and satisfying (23.2) is said to be *spectrally invariant*.)

Bibliography

Agmon, S., Douglis, A. and Nirenberg, L. (1959) Estimates near the boundary for solutions of elliptic partial differential equations satisfying general boundary conditions, I, *Comm. Pure Appl. Math.* **12**, pp.623–727.

Arveson, W. (2002) *A Short Course on Spectral Theory* (Springer).

Atiyah, M. F. (1997) The index of elliptic operators, in *Fields Medallists' Lectures*, World Scientific and Singapore University Press, pp.115–127.

Atkinson, F. V. (1951) The normal solvability of linear equations in normed spaces, *Mat. Sbornik N.S.* **28**, pp.3–14.

Brislawn, C. (1988) Kernels of trace class operators, *Proc. Amer. Math. Soc.* **104**, pp.1181–1190.

Dasgupta, A. (2008) *The Twisted Laplacian, the Laplacians on the Heisenberg Group and SG Pseudo-Differential Operators*, Ph.D. Dissertation, York University.

Dasgupta, A. (2008) Ellipticity of Fredholm pseudo-differential operators on $L^p(\mathbb{R}^n)$, in *New Developments in Pseudo-Differential Operators*, Operator Theory: Advances and Applications **189**, Birkhäuser, pp.107–116.

Dasgupta, A. and Wong, M. W. (2008) Spectral theory of SG pseudo-differential operators, *Studia Math.* **187**, pp.186–197.

Dasgupta, A. and Wong, M. W. (2010) Spectral invariance of SG pseudo-differential operators on $L^p(\mathbb{R}^n)$, in *Pseudo-Differential Operators: Complex Analysis and Partial Differential Equations*, Operator Theory: Advances and Applications **205**, Birkhäuser, pp.51–57.

Davies, E. B. (1980) *One-Parameter Semigroups* (Oxford University Press).

Davies, E. B. (2007) *Linear Operators and their Spectra* (Cambridge University Press).

Douglas, R. G. (1998) *Banach Algebra Techniques in Operator Theory*, 2nd edn. (Springer).

Fan, Q. and Wong, M. W. (1997) A characterization of Fredholm pseudo-differential operators, *J. London Math. Soc.* (2) **55**, pp.139–145.

Fan, Q. and Wong, M. W. (1997) A characterization of some non-elliptic pseudo-differential operators as Fredholm operators, *Forum Math.* **9**, pp.17–28.

Friedman, A. (1969) *Partial Differential Equations* (Holt, Rinehart and Winston).

Friedman, A. (1971) *Advanced Calculus* (Holt, Rinehart and Winston).

Gårding, L. (1953) Dirichlet's problem for linear partial differential equations, *Math. Scand.* **1**, pp.55–72.

Gel'fand, I. M. (1960) On elliptic equations, *Russian Math. Surveys* **15**, pp.113–123.

Grushin, V. V. (1970) Pseudo-differential operators on $\mathbb{R}^n$ with bounded symbols, *Funct. Anal. Appl.* **4**, pp.202–212.

Harutyunyan, G. and Schulze, B.-W. (2008) *Elliptic Mixed, Transmission and Singular Crack Problems* (European Mathematical Society).

Hess, P. and Kato, T. (1970) Peturbation of closed operators and their adjoints, *Comment. Math. Helv.* **45**, pp.524–529.

Hille, E. and Phillips, R. S. (1974) *Functional Analysis and Semi-Groups*, 3rd printing of revised (1957) edn. (American Mathematical Society).

Hörmander, L. (1960) Estimates for translation invariant operators in L^p spaces, *Acta Math.* **104**, pp.93–140.

Iancu, G. M (1999) *Dynamical Systems Generated by Pseudo-Differential Operators*, Ph.D. Dissertation, York University.

Kumano-go, H. (1981) *Pseudo-Differential Operators* (MIT Press).

Lax, P. D. (2002) *Functional Analysis* (Wiley-Interscience).

Reed, M. and Simon, B. (1980) *Functional Analysis,* revised and enlarged edn. (Academic Press).

Royden, H. L. (1988) *Real Analysis*, 3rd edn. (Prentice-Hall).

Rudin, W. (1987) *Real and Complex Analysis*, 3rd edn. (McGraw-Hill).

Saint Raymond, X. (1991) *Elementary Introduction to the Theory of Pseudodifferential Operators* (CRC Press).

Schechter, M. (1977) *Modern Methods in Partial Differential Equations* (McGraw-Hill).

Schechter, M. (1986) *Spectra of Partial Differential Operators*, 2nd edn. (North-Holland).

Schechter, M. (2002) *Principles of Functional Analysis*, 2nd edn. (American Mathematical Society).

Schulze, B.-W. (1998) *Boundary Value Problems and Singular Pseudo-Differential Operators* (Wiley).

Wheeden, R. L. and Zygmund A. (1977) *Measure and Integral* (Marcel Dekker).

Wong, M. W. (1983) Fredholm pseudo-differential operators on weighted Sobolev spaces, *Ark. Mat.* **21**, pp.271–282.

Wong, M. W. (1987) On some spectral properties of elliptic pseudo-differential operators, *Proc. Amer. Math. Soc.* **99**, pp.683–689.

Wong, M. W. (1988) Essential spectra of elliptic pseudo-differential operators, *Comm. Partial Differential Equations* **13**, pp.1209–1221.

Wong, M. W. (1991) Minimal and maximal operator theory with applications, *Canad. J. Math.* **43**, pp.617–627.

Wong, M. W. (1994) Spectral theory of pseudo-differential operators, *Adv. in Appl. Math.* **15**, pp.437–451.

Yosida, K. (1995) *Functional Analysis*, reprint of the 6th (1980) edn. (Springer-Verlag).

Index

Printed in the United States
By Bookmasters